Evolutionary Economics and Social Complexity Science

Volume 25

Mieko Tanaka-Yamawaki • Yumihiko Ikura

Principal Component Analysis and Randomness Test for Big Data Analysis

Practical Applications of RMT-Based Technique

Springer

Mieko Tanaka-Yamawaki
Organization for the Strategic Coordination of Research and Intellectual Properties (OSRI)
Meiji University
Tokyo, Japan

Yumihiko Ikura
Department of Mathematical Sciences Based on Modeling and Analysis, School of Interdisciplinary Mathematical Sciences
Meiji University
Tokyo, Japan

ISSN 2198-4204 ISSN 2198-4212 (electronic)
Evolutionary Economics and Social Complexity Science
ISBN 978-981-19-3969-3 ISBN 978-981-19-3967-9 (eBook)
https://doi.org/10.1007/978-981-19-3967-9

This Springer imprint is published by the registered company Springer Nature Singapore Pte Ltd.
The registered company address is: 152 Beach Road, #21-01/04 Gateway East, Singapore 189721, Singapore

Preface

This book was written to demonstrate the concept and usefulness of random matrix theory (RMT) in big data analysis, with emphasis on two RMT-oriented methodologies, RMT-PCA and RMT-test. Both are algorithms used in high-speed computer works. The book provides a thorough explanation of the methodologies and the output of real data, along with examples of programs and their outputs. Care is taken to convey the details of the methodology to potential readers who wish to analyze big data; to extract principal components using RMT-PCA; or to classify data by measuring the degree of randomness using RMT-test.

The data structure for applying RMT-PCA is limited to a rectangular matrix, having N rows (vertical) $\times$ L columns (horizontal), in which the element at the i-th row $\times$ the j-th column contains the value of the i-th quantity recorded at the j-th time. In other words, the i-th row contains time series of length L for the i-th quantity recorded at L prescribed times. That can be viewed as an N-dimensional vertical vector whose elements are L-dimensional horizontal vectors having equal-time stamps so that an inner product between any two rows represents an equal time correlation of the chosen rows.

This may sound too rigid. However, data that satisfy these conditions either exist as is, or can be made by pre-processing given data in order to satisfy the condition. In addition, data collection has become more active in recent years, and a vast amount of data has been accumulated, most of which are now regularly recorded at defined intervals of time: daily, hourly, and secondly. At the time when many of the studies documented in this book were actually conducted, data that met the requirements were still scarce and difficult to obtain. As to the price data used in Chap. 3, only three sets of price data were at our hands: (1) the daily-close stock prices of NYSE downloaded from web pages; (2) tickwise stock prices in the period of 1993–2002 purchased with the help of a Grant-in-Aid for Scientific Research; and (3) 1-minute-frequency stock prices of TSE in the period of 2007–2009 shared by Professor Tetsuya Takaishi of Hiroshima University of Economics. Using those three kinds of stock data, empirical experiments were conducted. The daily close prices are N-by-L rectangular matrices and can be readily used in RMT-PCA. However, the intraday prices need to be processed to satisfy the required shape. Tick-by-tick price data is labor intensive. It may require tedious preprocessing of the data before applying the algorithm. In practice, it is prudent to select data that do not require preprocessing.

The application of RMT-PCA is not limited to the stock market. Any suitable data set acquired at a given point in time can also be fed into RMT-PCA. A wide range of applications are expected, including weather, global economy, and medical diagnosis. For example, in medical diagnosis, it is possible to extract information about a patient's condition using an N-by-L matrix of N time-series physiological data representing the patient's condition, such as body temperature, blood pressure, and heart rate.

The second of the two topics addressed in this book, RMT-test, was developed by the authors to measure the degree of randomness of a given one-dimensional array. Although the methodology is partially parallel to RMT-PCA, the process of constructing an N-by-L rectangular matrix to create a correlation matrix and comparing its eigenvalue distribution with the RMT distribution completely removes the restriction on data, especially on data with equal-time correlation, allowing almost any big data to be used for analysis. Therefore, the applicability of RMT-test is infinitely broader than the original RMT-PCA, making it suitable for big data analysis. Several applications are presented in this book, including comparing hash functions, selecting stocks, and predicting stock index crashes.

The two methods presented here, which use the distribution of eigenvalues of correlation matrices created from long time series with strong randomness, also offer a comprehensive view into a world governed by simple and beautiful generality that transcends differences in detailed properties.

This book is the result of collaborative research conducted in the latter half of the year with young people fascinated by such beautiful generality at the Graduate School of Engineering, Tottori University, where one of the authors (MTY) was enrolled from 2003 to 2016. Some of their computational results are presented here. In particular, the results of Takemasa Kido, Yuko Tanaka, and Atsushi Yamamoto, who worked together on RMT-PCA in the first half, and Yuta Mikamori and Xin Yang, who worked together on RMT-test in the second half, are presented here with Figures and tables borrowed from these authors.

The results in Sect. 5.3 are based on the most recent results obtained by us (MTY and YI) in the period of 2018–2022, in order to add a new scope to the application of RMT-test.

We would be very pleased if these results correctly convey our intention to present a ready-to-use approach for big data analysis.

Tokyo, Japan

Mieko Tanaka-Yamawaki
Yumihiko Ikura

Contents

Chapter 1
Big Data Analysis with RMT

The purpose of this book is to introduce the basic concepts of RMT-oriented methods and their practical applications in big data analysis, focusing on two topics, RMT-PCA and RMT-test. Both are methodologies for analyzing large numerical data via computer programming. The essence of these methodologies is to use RMT to subtract random portions from the data under analysis and extract a small number of useful elements. RMT itself has a long history (Wigner 1955; Mehta 2004; Edelman and Rao 2005) and has a broad scope that is not limited to any particular method (e.g., Bahcall 1996, Beenakker 1997, Franchini and FKravtsov 2009, Peyrache et al. 2010). We focus on the above two methods and their applications and pay less attention to the other scope.

The first topic is principal component analysis (PCA), which we name "RMT-PCA." Chapters 2 and 3 are devoted to this topic. Chapter 2 summarizes the methodology, and Chap. 3 deals with some applications related to the selection of active stock industry sectors as principal components in each period. The main objective of RMT-PCA is to extract a small number of important factors from a large amount of numerical data. This is a tremendous task, similar to searching for a sunken treasure ship in the ocean. In this book, using high-frequency stock price data as a concrete example of data, we analyze the high-frequency stock prices of a large number of stocks and attempt to identify and extract the most actively traded industrious sectors in the stock market in each period using PCA. Of course, it seems that the most active stock issues can be easily identified by comparing the size of transactions. Why, then, is it necessary to analyze the price time series of a vast number of high-frequency trading records and extract from them a subset of stocks that play a major role?

Our ultimate objective is to establish the RMT-PCA methodology and demonstrate its effectiveness in the actual stock market. To do so, we can use the results obtained by RMT-PCA as material and verify whether the results are consistent with those obtained using the number of trades and the size of funds. A sudden increase in the trading volume of a stock can have different meanings depending on whether

M. Tanaka-Yamawaki, Y. Ikura, *Principal Component Analysis and Randomness Test for Big Data Analysis*, Evolutionary Economics and Social Complexity Science 25,
https://doi.org/10.1007/978-981-19-3967-9_1

it is the result of the actions of a few huge institutions or the aggregation of the decisions of a large number of individual investors. Such detailed and multifaceted analysis can sometimes be read into the stock price movements of individual stocks.

Considering this, we believe that a variety of factors are already incorporated into stock price movements and that by applying numerical calculations using RMT to stock price data, the essence of the data can be read, and the full potential of big data analysis can be demonstrated in classifying and labeling the characteristics of each piece of data. This is the reason why the method presented in this paper is so effective.

The effectiveness of the methods introduced here will thereby extend not only to the stock market but also to a wider range of fields, such as weather forecasting and climate change, which are thought to be governed by much more complex laws than stock prices, the use of biological information for diagnosis, and earthquake prediction. We chose high-frequency stock prices as the first application because financial data are readily available and the events of rising and falling stock prices are relatively simple.

The second topic is a method for measuring the degree of randomness of a set of numerical data, which we name the "RMT-test." Chapters 4 and 5 are devoted to this topic. The purpose is to introduce a useful tool for measuring the degree of randomness and to show the hope that such a tool can be used to label the properties of bulk data. But why do we care so much about randomness? One reason is that good tools to quantify randomness are hard to find, just the same as the difficulty finding good random generators which are hard to find (Park and Miller 1988) as well. But this is not the whole story.

It is widely believed that price fluctuations are random and that their probability distribution follows a normal (Gaussian) distribution. On the other hand, the probability distribution of price fluctuations is said to be long-tailed. How can we reconcile these two opposing ideas? The long tail means that the probability of large fluctuations is higher than that of a normal distribution, but even in the case of a large decline over several days, for example, the movements in smaller time frames are more violent and not just one-way decline, as can be seen in the following example. If we increase the resolution of the time interval to daily, hourly, minutely, or secondly, we can see that a different price chart is drawn, and accordingly, the probability distribution of fluctuation itself is also different.

The high randomness and the statistical distribution are two independent characteristics, but they both depend on the time resolution, so that what appears random at low resolution becomes correlated when the resolution is increased to a certain degree, and the correlation disappears when the resolution is increased further. For example, snow crystals may look beautifully symmetrical when viewed through a magnifying glass, but when shoveling heavy snow on either side of the road, they appear as nothing more than white lumps. In the same way, most animals and plants can be distinguished from each other only when they are viewed at a certain resolution, and at too coarse a resolution, they appear to be nothing more than a lump, while at too fine a resolution, they are nothing more than a collection of similar cells. The same is true of price fluctuations, and whether it is a chart or a probability distribution, it will change if you change the resolution, so the first prerequisite is to decide what kind of resolution you are talking about. In the case of price fluctuations, it is quite common for a day trader to follow the price on a 1-minute time frame and make a large profit in a few minutes, but a person who observes only the closing price for a week will see a completely different price fluctuation. With this in mind, in Chap. 5, Sect. 5.3, we will examine several 1-second stock price index movements and try to understand the market movements when they are close to random walks and when they are not, as two opposing states of the same market, i.e., random walk versus cascade transitions. The cascade state here corresponds to a long-tailed distribution with frequent large jumps in price movements, while the random walk state corresponds to a case where the 1-second price moves in a fine and random manner.

We focused on this problem and analyzed high-frequency price data. One reason to analyze high-frequency price data is that we need many data points for estimating the probability distribution from data. However, another reason which may be quite important is that the price movements and their statistical distribution strongly depend on its frequency.

Our recent analysis has shown that the statistical distribution of price differences per second does not always follow the same single distribution. Even if the distribution is long-tailed, the shape of the tail differs depending on the term and condition. In the case of a single stock price, the distribution approaches a normal distribution as the time interval between ticks approaches shorter than a few seconds. The statistical distribution of stock index prices, which are the weighted average of multiple stocks per second, exhibits long-tailed distributions of various types (Tanaka-Yamawaki et al. 2018; Tanaka-Yamawaki and Yamanaka 2019). What is needed is to find the conditions for a particular distribution to be realized. Looking at the level of randomness of price fluctuations, the normal distribution should have the highest randomness. If so, the degree of randomness of the price fluctuations can give us an idea of the market conditions. If the randomness is high and the statistical distribution of price fluctuations follows a normal distribution for a while, we would expect some anomaly to occur if the level of randomness suddenly decreases. In fact, a similar situation occurred in 2021, with price fluctuations basically following a normal distribution during the spring and summer of 2021, followed by a sudden drop in randomness during the summer and fall, reflecting the unstable psychology of investors who fear falling stock prices.

In other words, the high randomness of price fluctuations can be a strong candidate to label each data set. Usually, when we are confronted with big data, the first thing we do is to examine the various characteristics that characterize the data set, such as data length, data type, and when and where the data was collected. Once the formal questions are answered, the next step is to find the key characteristics of the data. Especially for financial time series that are highly random but contains information hidden under the randomness, we look for a good tool to remove randomness from the contaminated data and then to reduce the original dimension to find out the essential part of the data described in much smaller dimension. In other words, we first attempt to extract the principal component of the data. This is

equation is very small, less than 1%. However, the sample mean of 100 samples was found to be only a few percent due to large fluctuations.

We also applied the same tool to log-return columns of real-world data, e.g., LCG, MT, and PRN, and proved that the process of taking log-return of the original time series adds a certain "off-randomness" to the data sequence, which reduces the randomness of such log-return columns. The average difference between measured moments and their ideal values can be as high as 10–20%, and the lack of good quality data with intermediate levels of randomness that meet the RMT requirements has made it difficult to bridge the gap between "highly random" and "not random" in real-world data. Therefore, to clarify the boundary between the two extreme data, a series of arrays with different randomness levels are artificially prepared.

By using those artificial data having various levels of randomness, we have done a comparison of the RMT-test with the standard NIST randomness test. We have found that the shuffling reaches a saturation point at around 1.7 million shuffles, beyond which both NIST-test and our RMT-test exhibit no further improvement of randomness.

We show three kinds of practical applications of the RMT-test. The first application is the comparison of two Hash functions, by comparing the randomness levels of outputs of two Hash functions: MD5 and Sha-1. Naturally, outputs of the later Sha-1 are expected to have higher randomness level compared to the outputs of the older MD5. This expectation has shown to be proved.

The second application is to check the possible relationship of the randomness level of high-frequency price fluctuations of individual stocks and its future performance. By using stock prices of 1-minute resolution in 2007, 2008, and 2009, we have extracted a rule "stock of higher randomness performs better than the stock of lower randomness in the following term." This rule is supported by a strong correlation between the degree of randomness computed from the 1-minute price fluctuations of each year and the behavior of daily prices in the following year. The result is supported by the data. The performance of the stock of the highest/lowest randomness performed better/worse in the following year compared to the average stock prices, for all of the 3 years in 2007–2009, excluding the terms of Lehman shock. However, these 3 years are the period of very week stock market, in which almost all the stock prices fell in this period. For this reason, we tried to examine what happens in the rising market. Applying the same on rising market in the USA in the years of 1993–1996, we checked whether the most/least random stock perform better/worse than other stocks in the following year for each year in the period of 1993–1996. The result was not clear, but 3 years out of 4, the assumption seems to hold.

In the third and the final application, we measured the randomness levels of various stock indices at various time intervals and showed that a sudden decrease of randomness level indicates a possible market decline, when the market is in the mode of random walk. The contents of the remaining part of this book are as follows. In Chap. 2, we provide a step-by-step introduction to the methodology, keeping in mind its theoretical background in order to develop a formulation common to both RMT-PCA and RMT-test. The differences between the two are as follows: RMT-

PCA deals with price time series of N stocks to construct an N-by-L rectangular matrix, while RMT-test constructs an N-by-L rectangular matrix by slicing a single price time series into N pieces of length L. Thus, the correlation matrix C to be diagonalized is the correlation between different pairs of stocks in RMT-PCA, while the same C in RMT-test is the correlation between two different parts in a single price time series of length N times L. In other words, C in RMT-PCA is the cross-correlation between pairs of stocks, while C in RMT-test is the auto-correlation between pairs of different parts of the same stock or index time series, etc. After constructing C, the diagonalization process to find eigenvalues can be handled by Jacobi rotation due to the symmetry of the matrix C. In Chap. 3, the RMT-PCA method is applied to tick-by-tick price time series of the Japanese and US stock markets. Chapter 4 describes the formulation and application of the RMT-test, both qualitative and quantitative versions, and how the parameters N, Q, and k are determined, using various random number generators. We also compare the RMT-test to NIST, a standard randomness test, by artificially constructing random number sequences with various randomness levels and show that the RMT-test is as capable as NIST. We also show an example of applying the RMT-test to the comparison of two hash functions and measuring the degree of randomness of their outputs. Although some parts of the beginning of Chap. 4 overlap with Chap. 2, we dare allowed the overlap, in order to avoid referring to the previous section in the middle of the paper. Finally, in Chap. 5, we examine the possibility of predicting future stock prices using the randomness measured by the RMT-test. In Sect. 5.2, we measure the randomness of individual stock prices by year and attempt to predict the performance in the following year. In Sect. 5.3, we applied the method on the time variation of randomness of stock index such as TOPIX and TOPIXcore30 in 1-minute resolution and considered the possibility of predicting a market decline. Chapter 6 is devoted to conclusion.

Chapter 2
Formulation of RMT-PCA

2.1 From Data to Rectangular Matrix

When dealing with big data, it is stored in computers as digital data according to some regularity. A familiar example is stock prices. Usually, the stock prices of each company are arranged in a time series. A vector is a mathematical concept that is useful for handling such a long array of stock prices in a single character. For example, by defining a vector, referring to all price sets of size L by a single word $\boldsymbol{g}_i$, we have

$$\boldsymbol{g}_i = (g_{i,1}\ g_{i,2}\ \cdots\ g_{i,L}) \tag{2.1}$$

We can also define the "transverse" of this vector by changing the direction of the array: from horizontal array to the vertical array, such as

$$\boldsymbol{g}_i^t = \begin{pmatrix} g_{i,1} \\ g_{i,2} \\ \vdots \\ g_{i,L} \end{pmatrix} \tag{2-1'}$$

Here we have used a suffix i for the sake of numbering the company to count N companies by $i = 1, \ldots, N$. To go further, it would be more convenient to introduce a mathematical notion of matrix, that is, a vector of a vector, such as

$$G = \begin{pmatrix} \boldsymbol{g}_1 \\ \boldsymbol{g}_2 \\ \vdots \\ \boldsymbol{g}_N \end{pmatrix} = \begin{pmatrix} g_{1,1} & g_{1,2} & \cdots & g_{1,L} \\ g_{2,1} & g_{2,2} & \cdots & g_{2,L} \\ \vdots & \vdots & \ddots & \vdots \\ g_{N,1} & g_{N,2} & \cdots & g_{N,L} \end{pmatrix} \tag{2.2}$$

M. Tanaka-Yamawaki, Y. Ikura, *Principal Component Analysis and Randomness Test for Big Data Analysis*, Evolutionary Economics and Social Complexity Science 25,
https://doi.org/10.1007/978-981-19-3967-9_2

G^t is the transverse matrix of the N-by-L matrix.

$$G^t = \begin{pmatrix} g_{1,1} & g_{2,1} & \cdots & g_{N,1} \\ g_{1,2} & g_{2,2} & \cdots & g_{N,2} \\ \vdots & \vdots & \ddots & \vdots \\ g_{1,L} & g_{2,L} & \cdots & g_{N,L} \end{pmatrix} \tag{2.3}$$

By defining G, a huge data set of size N-by-L can be referred to by a single word G, where the small letter t on the shoulder of the letter means transverse, i.e., changing the way to represent an array, from horizontal to vertical or from vertical to horizontal.

Thousands of stocks are traded in the stock market every day. Some of them are heavily traded, while some others are not. For example, if we focus on stocks such as the Nikkei 225 Stock Average, Dow Jones Industrial Average, S&P500, or TOPIX, can we somehow extract the major sectors every hour? If we could design an AI program that could analyze all the trading data on the net and tell us which stocks are good or which ones to avoid, market analysis would be an exciting scientific game. We shall try to construct such a system using Random Matrix Theory (RMT) as follows. Given a record of all trades in every hour, and in particular the prices of N (= 500 for the S&P500) stocks, we can extract some leading stocks by the method described in this chapter.

Consider a set of L prices for N stocks, put them into the form of G in Eq. (2.2). For example, $i = 1$ for Microsoft, $i = 2$ for Apple Computer, etc., that is

$$G = \begin{pmatrix} \text{Msoft}_1 & \text{Msoft}_2 & \cdots & \text{Msoft}_L \\ \text{Apple}_1 & \text{Apple}_2 & \cdots & \text{Apple}_L \\ \vdots & \vdots & \ddots & \vdots \\ \text{stockN}_1 & \text{stockN}_2 & \cdots & \text{stockN}_L \end{pmatrix} \tag{2-2'}$$

This matrix G contains all the price information for N companies for L days (minutes, or seconds, etc.). For readers unfamiliar with tools such as vectors, matrices, and inner product, see Appendix A.

Note, however, that the price increment is more important than the stock price. The price increment is the difference between the current price and the previous price, indicating whether the price has gone up or down. However, this increment is still inconvenient for comparing different stocks. This is because the order of price of stock A and the order of price of stock B may be very different, in which case it would not make sense to compare the increments of the two stock prices. A more useful value is the rate of increase or decrease, called "return." For example, if the price goes up 10%, the return is 0.1, and if it goes down 20%, the return is −0.2, and so on. Furthermore, log-difference is often used to compute return, as we come back to this issue in Chap. 3. The return has no unit, unlike the price depends on the unit of currency that depends on each nation. From this point forward, any discussion of price implies return. Moreover, this vector in Eq. (2.1) is a zero-mean vector for all $i = 1, \ldots, N$, and is assumed to be normalized to zero mean and unit norm (Appendix A), such that for all $i = 1, \ldots, N$,

$$\sum_{k=1}^{L} g_{i,k} = 0 \tag{2.4}$$

$$|\boldsymbol{g}_i| = \sqrt{\sum_{k=1}^{L} (g_{i,k})^2} = 1 \tag{2.5}$$

2.2 Correlation Matrices and Their Properties

The cross-correlation matrix C is constructed by multiplying the rectangular matrix G and its transverse matrix G^t as follows:

$$C = GG^t \tag{2.6}$$

By components, this is written as

$$C = \begin{pmatrix} c_{1,1} & c_{1,2} & \cdots & c_{1,N} \\ c_{2,1} & c_{2,2} & \cdots & c_{2,N} \\ \vdots & \vdots & \ddots & \vdots \\ c_{N,1} & c_{N,2} & \cdots & c_{N,N} \end{pmatrix} = \begin{pmatrix} \boldsymbol{g}_1 \cdot \boldsymbol{g}_1 & \boldsymbol{g}_1 \cdot \boldsymbol{g}_2 & \cdots & \boldsymbol{g}_1 \cdot \boldsymbol{g}_N \\ \boldsymbol{g}_2 \cdot \boldsymbol{g}_1 & \boldsymbol{g}_2 \cdot \boldsymbol{g}_2 & \cdots & \boldsymbol{g}_2 \cdot \boldsymbol{g}_N \\ \vdots & \vdots & \ddots & \vdots \\ \boldsymbol{g}_N \cdot \boldsymbol{g}_1 & \boldsymbol{g}_N \cdot \boldsymbol{g}_2 & \cdots & \boldsymbol{g}_N \cdot \boldsymbol{g}_N \end{pmatrix} \tag{2.7}$$

where each element in column j by row i of the correlation matrix C is the inner product of the i-th vector $\boldsymbol{g}_i$ and the j-th vector $\boldsymbol{g}_j$ defined as

$$\boldsymbol{g}_i \cdot \boldsymbol{g}_j = \sum_{k=1}^{L} g_{i,k} g_{j,k} \tag{2.8}$$

The element of the i-th row and the j-th column of matrix C is denoted as the (i, j) component of C and is written as $C_{i,j}$. By the definitions in Eq. (2.7), C is symmetric about the exchange of i (row) and j (column), and we have

$$C_{i,j} = C_{j,i} \ (i = 1, \ldots, N; \ j = 1, \ldots, N) \tag{2.9}$$

Also, by the normalization condition in Eq. (2.5), all diagonal elements are equal to 1.

$$C_{i,i} = 1 \ (i = 1, \ldots, N) \tag{2.10}$$

Furthermore, the absolute values of its off-diagonal elements are generally less than 1.

$$|C_{i,j}| \leq 1 \ (i = 1, \ldots, N; \ j = 1, \ldots, N) \tag{2.11}$$

Recall that the inner product of two vectors $\boldsymbol{g}_i$ and $\boldsymbol{g}_j$ is defined as $\boldsymbol{g}_i \cdot \boldsymbol{g}_j = |\boldsymbol{g}_i||\boldsymbol{g}_j| \cos\theta$, where θ is the angle between $\boldsymbol{g}_i$ and $\boldsymbol{g}_j$. In the case of unit vectors, $|\boldsymbol{g}_i| = |\boldsymbol{g}_j| = 1$, so the magnitude of the inner product is $|\cos\theta| \leq 1$. From this definition, each element $C_{i,j}$ of the correlation matrix C represents the degree of overlap between the two price time series $\boldsymbol{g}_i$ and $\boldsymbol{g}_j$.

Once the correlation matrix C is established, its eigenvalues are then determined and compared to the theoretical distribution derived by the RMT.

2.3 Eigenvalues of a Correlation Matrix

As is well known, any real symmetric matrix C can be diagonalized by the similarity transformation $V^{-1}CV = V^tCV$ by an orthogonal matrix V satisfying $V^{-1} = V^t$, where V^{-1} and V^t represent the inverse matrix and the transverse matrix of matrix V, respectively. This matrix V can be expressed by using the eigenvectors of the matrix C

$$V = \begin{pmatrix} v_{1,1} & v_{2,1} & \cdots & v_{N,1} \\ v_{1,2} & v_{2,2} & \cdots & v_{N,2} \\ \vdots & \vdots & \ddots & \vdots \\ v_{1,N} & v_{2,N} & \cdots & v_{N,N} \end{pmatrix} \tag{2.12}$$

where the k-th column of V is the k-th eigenvector $\boldsymbol{v}_k$ of a matrix C, such as

$$\boldsymbol{v}_k = \begin{pmatrix} v_{k,1} \\ v_{k,2} \\ \vdots \\ v_{k,N} \end{pmatrix} \tag{2.13}$$

The eigenvalues λ_k and the associated eigenvectors $\boldsymbol{v}_k$ of a matrix C satisfy the following equation

$$C\boldsymbol{v}_k = \lambda_k \boldsymbol{v}_k \ (k = 1, \ldots, N) \tag{2.14}$$

Or, equivalently by writing all the components explicitly,

$$\begin{pmatrix} c_{1,1} & c_{1,2} & \cdots & c_{1,N} \\ c_{2,1} & c_{2,2} & \cdots & c_{2,N} \\ \vdots & \vdots & \ddots & \vdots \\ c_{N,1} & c_{N,2} & \cdots & c_{N,N} \end{pmatrix} \begin{pmatrix} v_{k,1} \\ v_{k,2} \\ \vdots \\ v_{k,N} \end{pmatrix} = \lambda_k \begin{pmatrix} v_{k,1} \\ v_{k,2} \\ \vdots \\ v_{k,N} \end{pmatrix} \tag{2.15}$$

By normalizing $\boldsymbol{v}_k$ such that

$$\boldsymbol{v}_k \cdot \boldsymbol{v}_k = \sum_{n=1}^{N} v_{k,n}^2 = 1 \tag{2.16}$$

the eigenvalue equation can also be written as

$$\boldsymbol{v}_l^t C \boldsymbol{v}_k = \lambda_k \delta_{k,l} \tag{2.17}$$

where $\delta_{k,l}$ is 0 for $k \neq l$ and is 1 for $k = l$. This $\delta_{k,l}$ is called as Kronecker's delta. It can also be written explicitly with components as follows:

$$\begin{pmatrix} v_{l,1} v_{l,2} \cdots v_{l,N} \end{pmatrix} \begin{pmatrix} c_{1,1} & c_{1,2} & \cdots & c_{1,N} \\ c_{2,1} & c_{2,2} & \cdots & c_{2,N} \\ \vdots & \vdots & \ddots & \vdots \\ c_{N,1} & c_{N,2} & \cdots & c_{N,N} \end{pmatrix} \begin{pmatrix} v_{k,1} \\ v_{k,2} \\ \vdots \\ v_{k,N} \end{pmatrix} = \lambda_k \delta_{k,l} \tag{2.18}$$

Since the eigenvectors with different suffices k and l are orthogonal to each other,

$$\boldsymbol{v}_k \cdot \boldsymbol{v}_l = \sum_{n=1}^{N} v_{k,n} v_{l,n} = 0 \tag{2.19}$$

Using Kronecker's delta, Eqs. (2.16) and (2.19) can be written as

$$\boldsymbol{v}_k \cdot \boldsymbol{v}_l = \delta_{k,l} \tag{2.20}$$

In matrix representation, the above process can be written as

$$
\begin{aligned}
&V^t C V\\
&= \begin{pmatrix} v_{1,1} & v_{1,2} & \cdots & v_{1,N} \\ v_{2,1} & v_{2,2} & \cdots & v_{2,N} \\ \vdots & \vdots & \ddots & \vdots \\ v_{N,1} & v_{N,2} & \cdots & v_{N,N} \end{pmatrix} \begin{pmatrix} 1 & c_{1,2} & \cdots & c_{1,N} \\ c_{2,1} & 1 & \cdots & c_{2,N} \\ \vdots & \vdots & \ddots & \vdots \\ c_{N,1} & c_{N,2} & \cdots & 1 \end{pmatrix} \begin{pmatrix} v_{1,1} & v_{2,1} & \cdots & v_{N,1} \\ v_{1,2} & v_{2,2} & \cdots & v_{N,2} \\ \vdots & \vdots & \ddots & \vdots \\ v_{1,N} & v_{2,N} & \cdots & v_{N,N} \end{pmatrix}\\
&= \begin{pmatrix} \lambda_1 & 0 & \cdots & 0 \\ 0 & \lambda_2 & \cdots & 0 \\ \vdots & \vdots & \ddots & \vdots \\ 0 & 0 & \cdots & \lambda_N \end{pmatrix}
\end{aligned}
\tag{2.21}
$$

is represented by the matrix V with the k-th eigenvector of matrix C as the k-th column when $\lambda_k \neq \lambda_l$ for $k \neq l$.

In fact, the eigenvalues can be obtained by diagonalizing the matrix C by repeating rotations according to Jacobi's diagonalization algorithm, as long as the matrix C is symmetric. That is, each rotation is performed in the following three steps:

(1) Find the largest (in absolute value) off-diagonal element $c_{k,l}$ such that $|c_{k,l}| = \max\{c_{i,j}\}$.
(2) Apply the rotation matrix $W(\theta)$ to the k-th row and the l-th column, such that

$$
\begin{aligned}
&W_{k,k} = \cos\theta = W_{l,l} \text{ and } W_{k,l} = -\sin\theta = -W_{l,k}\\
&W_{i,i} = 1 \text{ for } i \neq k, l\\
&W_{i,j} = 0 \text{ for } i \neq j, \text{ and } i \neq k, l
\end{aligned}
$$

such that

$$
W(\theta) = \begin{pmatrix}
1\,0 & & & & & & 0\,0 \\
0\,1 & & & & & & 0\,0 \\
 & \ddots & & & & & \\
 & & \cos\theta & 0 \cdots 0 & -\sin\theta & & \\
 & & 0 & 1 & 0 & & \\
 & & \vdots & \ddots & \vdots & & \\
 & & 0 & 1 & 0 & & \\
 & & \sin\theta & 0 \cdots 0 & \cos\theta & & \\
 & & & & & \ddots & \\
0\,0 & & & & & & 1\,0 \\
0\,0 & & & & & & 0\,1
\end{pmatrix}
\tag{2.22}
$$

(3) Rotate C by W such that

$$\begin{pmatrix} \cos\theta & \sin\theta \\ -\sin\theta & \cos\theta \end{pmatrix} \begin{pmatrix} c_{i,i} & c_{i,j} \\ c_{i,j} & c_{j,j} \end{pmatrix} \begin{pmatrix} \cos\theta & -\sin\theta \\ \sin\theta & \cos\theta \end{pmatrix} = \begin{pmatrix} \tilde{c}_{i,i} & 0 \\ 0 & \tilde{c}_{j,j} \end{pmatrix} \tag{2.23}$$

After one step, $C \to C' = W^t C W$, $c'_{k,l} = 0$, and the absolute value of the largest matrix element of matrix C' is smaller than $|c_{k,l}|$.

The rotation is repeated by n times so that all the off-diagonal elements are smaller than some minimum number specified to be considered as zero.

$$\begin{aligned} C &\to W_1^t C W_1 \to W_2^t W_1^t C W_1 W_2 \to \cdots \to V^t C V \\ &= \begin{pmatrix} v_{1,1} & v_{1,2} & \cdots & v_{1,N} \\ v_{2,1} & v_{2,2} & \cdots & v_{2,N} \\ \vdots & \vdots & \ddots & \vdots \\ v_{N,1} & v_{N,2} & \cdots & v_{N,N} \end{pmatrix} \begin{pmatrix} 1 & c_{1,2} & \cdots & c_{1,N} \\ c_{2,1} & 1 & \cdots & c_{2,N} \\ \vdots & \vdots & \ddots & \vdots \\ c_{N,1} & c_{N,2} & \cdots & 1 \end{pmatrix} \begin{pmatrix} v_{1,1} & v_{2,1} & \cdots & v_{N,1} \\ v_{1,2} & v_{2,2} & \cdots & v_{N,2} \\ \vdots & \vdots & \ddots & \vdots \\ v_{1,N} & v_{2,N} & \cdots & v_{N,N} \end{pmatrix} \\ &= \begin{pmatrix} \lambda_1 & 0 & \cdots & 0 \\ 0 & \lambda_2 & \cdots & 0 \\ \vdots & \vdots & \ddots & \vdots \\ 0 & 0 & \cdots & \lambda_N \end{pmatrix} \end{aligned} \tag{2.24}$$

In this case, the eigenvectors are given by each column of the following diagonalizing matrix.

$$V = W_1 W_2 \cdots W_n \tag{2.25}$$

The details of this algorithm are summarized in Appendix B. Since the execution of this process is too time-consuming to be done manually, it is convenient to program it on a computer: a sample program written in C language is given in Appendix C for the interested reader's reference.

Note that, by the trace theorem, the sum of all eigenvalues is equal to N, the sum of the diagonal elements of the original matrix C. This means that some eigenvalues are greater than 1, indicating a strong correlation between those stocks. On the other hand, the remaining eigenvalues are smaller than 1, indicating little correlation between stocks in these categories.

After computing the eigenvalues of the correlation matrix made of the normalized price time series in this way, we put them in the descending order as follows:

$$\lambda_1 > \lambda_2 > \cdots > \lambda_N \tag{2.26}$$

Actually, according to RMT, the details of those values are not so important. If the data are totally random, the empirically obtained probability distribution (histogram) of those eigenvalues approaches a certain simple function, with parameter $Q = L/N$, the Marchenko-Pastur function. This is the essence of the data analyses with RMT-oriented methodology.

2.4 Eigenvalue Distribution and the RMT Formula

Historically, the most famous application of the RMT is Wigner's semi-circle law (Wigner 1955) where the distribution function $P(\lambda)$ of the eigenvalues of an N-by-N random matrix, whose element are the random numbers, approaches a semi-circle of radius determined by the standard deviation R, of the off-diagonal elements of that matrix, in the limit of N going to infinity, as follows:

$$P_{RMT}(\lambda) = \frac{1}{2\pi R^2}\sqrt{4R^2 - \lambda^2} \tag{2.27}$$

Note that all the details of the matrix elements have been washed out except for the standard deviation R. The resulting distribution depends only on one parameter R.

An application of RMT suitable for time series analysis is a correlation matrix whose elements are the inner products of random time series. Such a random matrix is called Wishart-type. The eigenvalue distribution of such a correlation matrix can be approximated to the Marchenko-Pastur distribution in the limit where N and L are going to infinity by keeping Q fixed.

$$P_{RMT}(\lambda) = \frac{Q}{2\pi}\frac{\sqrt{(\lambda_+ - \lambda)(\lambda - \lambda_-)}}{\lambda} \tag{2.28}$$

$$\lambda_\pm = 1 + \frac{1}{Q} \pm 2\sqrt{\frac{1}{Q}} \tag{2.29}$$

$$Q = \frac{L}{N} > 1 \tag{2.30}$$

Here, Q is a free parameter. The magnitude of the theoretical ranges $[\lambda_-, \lambda_+]$ for $Q = 2, 3, 4$ is summarized in Table 2.1.

Table 2.1 The $[\lambda_-, \lambda_+]$ for $Q = 2, 3, 4$

Q	λ_-	λ_+
2	0.085	2.914
3	0.179	2.488
4	0.250	2.250

Table 2.2 Three data sets used in Sect. 2.4 are tabulated

Name of data	Type	Market	Years
Data-A	tick-by-tick	New York	1993–2002
Data-B	day-by-day	New York	1994–2009
Data-C	minute-by-minute	Tokyo	2007–2009

The main purpose of this book is to apply Eqs. (2.28)–(2.30), in two ways: the first application is the RMT-PCA, and the second application is the RMT-test.

In RMT-PCA, eigenvalues larger than λ_+ and their eigenvectors are identified as principal components. The largest eigenvalue λ_1 and the associated eigenvector $\boldsymbol{v}_1$ are considered the first candidate. However, the component elements of the first eigenvector $\boldsymbol{v}_1$ form a flat spectrum, often called a market mode, and cannot indicate a specific signal. The next candidate for the principal components is the second largest eigenvalue λ_2 and its associated eigenvector $\boldsymbol{v}_2$. The components of the second eigenvector $\boldsymbol{v}_2$ have a structure that is deeply biased toward business sectors with strong stock prices.

This approach was shown in early works by Stanley and co-workers at Boston University, USA (Plerou et al. 1999, 2002) and Bouchaud and co-workers in France (Laloux et al. 1999). However, these earlier studies had relatively little data available at the time, and combining all of the price time series for several years in the 1990s yielded sufficiently large N and L. Stimulated by these earlier studies, one of the authors attempted a systematic analysis (Tanaka-Yamawaki et al. 2013), by using the three data sets; A, B, C shown in Table 2.2.

Using these three data sets, we aimed to pinpoint the extraction of trends in each period. In particular, for the tick-wise price data, we were able to obtain the major industry sectors from the clusters of elements of the second eigenvector corresponding to stocks in the same sector.

The second application is the RMT-test, which shows the methodology of Chap. 4 and its application to the various random sequences presented in Chap. 5.

2.5 RMT-PCA: RMT-Oriented Principal Component Analysis

The diagonalization process of the correlation matrix C by repeating Jacobi rotations is equivalent to transforming the set of normalized time series into a set of eigenvectors.

$$\boldsymbol{y}(t) = V\boldsymbol{x}(t) \tag{2.31}$$

By using components, this can be written explicitly as

$$y_i(t) = \sum_{j=1}^{N} v_{i,j} x_j(t) \tag{2.32}$$

The eigenvalues can be interpreted as the variance of the newly found variable found by rotating it toward the component with the largest variance among the N independent variables. In other words

$$\begin{aligned}\sigma^2 &= \frac{1}{T}\sum_{t=1}^{T} y_i^2(t) \\ &= \frac{1}{T}\sum_{t=1}^{T}\sum_{l=1}^{N} v_{i,l}x_l(t)\sum_{m=1}^{N} v_{i,m}x_m(t) \\ &= \sum_{l=1}^{N}\sum_{m=1}^{N} v_{i,l}v_{i,m}C_{l,m} \\ &= \lambda_i\end{aligned} \tag{2.33}$$

and from Eq. (2.4), the time average $\langle y_i \rangle$ of y_i is always zero. For simplicity, we name the eigenvalues in descending order, $\lambda_1 > \lambda_2 > \cdots > \lambda_N$.

The rationale for principal component analysis is to expect the magnitude of the principal component to stand out relative to the other components in N dimensional space. Figure 2.1 shows the image of principal components by presenting two-dimensional data (x, y) rotated to $z = ax + by$, where z is chosen to be the principal axis; this data set can be described as one-dimensional information along z.

If the magnitude of the largest eigenvalue λ_1 of C is significantly larger than the second largest eigenvalue λ_2, the data will be scattered primarily along this principal axis, corresponding to the direction of the eigenvector $\boldsymbol{v}_1$ of the largest eigenvalue λ_1.

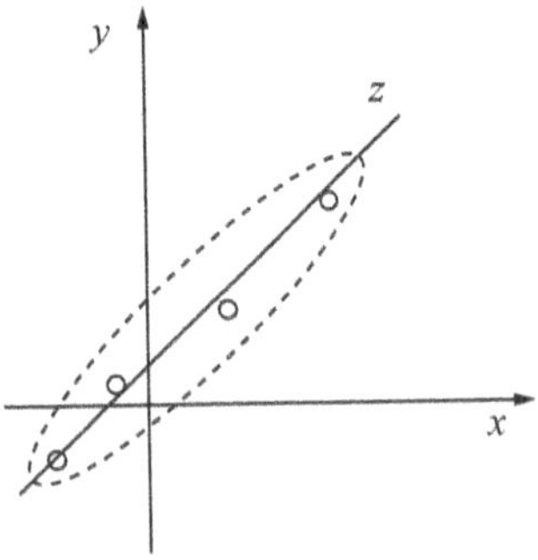

Fig. 2.1 Image of principal axis z along which most data points on (x, y) plane are located

This is the first principal component. Similarly, the second principal component can be identified as the eigenvector $\boldsymbol{v}_2$ perpendicular to $\boldsymbol{v}_1$, which corresponds to the second largest eigenvalue λ_2. When the dimension N is very large and the data are highly random, only a part of the large components carry the significant concentration, while the remaining components can be regarded as noise. In the case of a stock market, the number of stock N is usually larger than 100, and the length of the price time series L is also very long, easily satisfying the asymptotic conditions N and $L \to \infty$, when applying random matrix theory (hereafter, RMT). According to RMT, the eigenvalue distribution spectrum of C lies in the range between λ_- and λ_+

$$\lambda_- < \lambda < \lambda_+ \tag{2.34}$$

The criterion proposed in the RMT-PCA is to consider eigenvalues in the range, in Eq. (2.34), of the RMT spectrum as random noise and to consider as principal components those components whose eigenvalue λ_i is much larger than the upper RMT limit λ_+,

$$\lambda_+ \ll \lambda_i \tag{2.35}$$

Chapter 3
RMT-PCA for the Stock Markets

3.1 From Stock Prices to Log-Returns

We will examine how to extract a set of correlated stock prices from a large and complex network consisting of hundreds or thousands of stocks. In addition to correlations between stocks in the same industry, there are also correlations and anti-correlations between stocks in different industries. In order to compare price time series of different magnitude, profits are often used instead of prices. Profit (return) is defined as the ratio of increments ΔS to the price S at time t,

$$\frac{S(t+\Delta t) - S(t)}{S(t)} = \frac{\Delta S(t)}{S(t)} \tag{3.1}$$

A more convenient quantity, however, is the log-return, defined by the difference between log-prices

$$r(t) = \log S(t+\Delta t) - \log S(t) \tag{3.2}$$

Since we can also write

$$r(t) = \log \frac{S(t+\Delta t)}{S(t)} \tag{3.3}$$

and the numerator of log can be written as $S(t+\Delta t) = S(t) + \Delta S(t)$,

$$r(t) = \log\left(1 + \frac{\Delta S(t)}{S(t)}\right) \simeq \frac{\Delta S(t)}{S(t)} \tag{3.4}$$

It is essentially the same as the profit $r(t)$ defined in Eq. (3.1). The definition in Eq. (3.2) is more convenient to exhibit the additivity of this quantity rather than the

M. Tanaka-Yamawaki, Y. Ikura, *Principal Component Analysis and Randomness Test for Big Data Analysis*, Evolutionary Economics and Social Complexity Science 25, https://doi.org/10.1007/978-981-19-3967-9_3

ratio. Combined with Eq. (3.4), it can be seen that this quantity is a positive profit if $\Delta S > 0$ 0 and a loss if $\Delta S < 0$.

Equations (3.1)–(3.4) are written using continuous functions for continuous variables. Recall, however, that we need to use digital computer processing here. Change the continuous time variable t to a simple suffix $k = 1, \ldots, L$, and change the continuous functions $S_i(t)$ and $r_i(t)$ to discrete variables $S_{i,k}$ and $r_{i,k}$ to give their discrete versions of the relationship in Eq. (3.2)

$$r_{i,k} = \log S_{i,k+1} - \log S_{i,k} \tag{3.5}$$

where $S_{i,k}$ is the k-th price data of the i-th stock as provided by various databases.

For discrete variables, the cross-correlation matrix $C_{i,j}$ between the return of two stocks, i and j, can be written as the inner product of the two log-return time series,

$$C_{i,j} = \sum_{k=1}^{L} g_{i,k} g_{j,k} \tag{3.6}$$

where $g_{i,k}$ is the i-th time series normalized so that the mean is zero and the variances is one as follows:

$$g_{i,k} = \frac{r_{i,k} - \langle r_i \rangle}{\sigma_i} \qquad (i = 1, \ldots, N,\ k = 1, \ldots, L) \tag{3.7}$$

where

$$\langle r_i \rangle = \frac{1}{L} \sum_{k=1}^{L} r_{i,k} \tag{3.8}$$

and

$$\sigma_i^2 = \langle r_i^2 \rangle - \langle r_i \rangle^2 \tag{3.9}$$

3.2 The Methodology of the RMT-PCA

The role of RMT-PCA is to separate the signal from the noise. If stock prices really move randomly, the cross-correlation matrix C defined in Eq. (2.7), which consists of inner product between pairs of stock price time series of N stocks, should have eigenvalues that follow the RMT theoretical distribution function given by Eqs. (2.28)–(2.30). In reality, however, there is a strong correlation between the movements of many stocks, and once many stocks move together, the entire market tends to be affected by the wave. In addition, regardless of the financial condition of individual stocks, multiple stocks in the same industry often move in tandem, and observing changes in active industries is often an important piece of information that characterizes the market. Therefore, the actual distribution function obtained from the price time series is expected to deviate a random walk. The signal of this deviation can be confirmed by the presence of eigenvalues larger than the theoretical upper bound λ_+ of the Marchenko-Pastur distribution in Eq. (2.28) and away from the upper bound. Except for a few eigenvalues that exceed the bound λ_+, most eigenvalues computed from the RMT spectrum remain in the range $[\lambda_-, \lambda_+]$, if N is sufficiently large. Therefore, by picking up a few eigenvalues that exceed the boundary λ_+, the signal and noise can be separated.

The method uses the results of random matrix theory (RMT) as a principal component analysis of cross-correlation matrices of pairs of independent time series, applied on stock prices by a European group (Laloux et al. 1999) and a US group (Plerou et al. 1999) by combining daily closing prices of US stocks with other data and then applied to daily closing prices of Japanese stocks by Aoyama et al. (2010) and US group proposed and applied. If the price increments as the time series are random, the spectrum is expected to agree with the theoretically derived RMT formula (Sengupta and Mitra 1999), but in practice we observe principal components that deviate significantly from the random noise.

We collected and analyzed tick data for the US market from 1994 to 2002. This is Data-A in the list of the three data mentioned in Sect. 2.4. The higher frequency of data points makes us possible to compare results from different years and to describe historical changes in the market in the scenario of a new method of principal component analysis based on the RMT spectrum. For convenience, we refer to this method as RMT-PCA (Tanaka-Yamawaki et al. 2013).

Here $\boldsymbol{v}_2$ is used for the following reasons. First, it is known that the vector $\boldsymbol{v}_1$ has no dominant direction. In fact, in numerical experiments with multiple data sets, the N components of the eigenvector $\boldsymbol{v}_1$ corresponding to the largest eigenvalue $\boldsymbol{v}_3, \ldots, \boldsymbol{v}_n$, the only $\boldsymbol{v}_2$ carries information on the dominant directions.

This is the result of a calculation using a 1-minute chart of Japanese stocks (part of Data-C in Table 2.2). As will be explained in more detail later, Japanese stocks are numbered by industry, so it is easy to observe how stocks in the same industry are linked. $\boldsymbol{v}_1$, $\boldsymbol{v}_2$, $\boldsymbol{v}_3$, and $\boldsymbol{v}_4$ N-component bar charts are shown on the left. $\boldsymbol{v}_1$ is flat, $\boldsymbol{v}_2$ is biased toward certain industries, $\boldsymbol{v}_3$ is also biased, but the influence of the random component is quite strong. The right graph shows the probability density of occurrence of the corresponding left bar; the first line shows the probability density of occurrence of the value of the N component of $\boldsymbol{v}_1$, with values concentrated at 0.05. Contrasting the normalized distribution with the same standard deviation for comparison, we see that in $\boldsymbol{v}_4$, the values overlap into a nearly normal distribution. This leads us to infer that the clearest signal is obtained for $\boldsymbol{v}_2$. Therefore, we define RMT-PCA as the algorithm for tracking the time variation of dominant sectors by using only $\boldsymbol{v}_2$ and summarize it in Fig. 3.1.

Even so, if we assume from the beginning that we can track the time variation of dominant sectors by using only $\boldsymbol{v}_2$, we may miss some important information. We should investigate what these contributions could be. First, we will trace the time variation of each dominant sector using the principal components from $\boldsymbol{v}_2$ to

Algorithm of RMT-PCA:

1. Select N stock codes that have at least one transaction in all time stamps of $k = 1, \ldots, L+1$.
2. Compute the log-return $r_i(t_k)$ for all N stocks selected. Normalize the time series t_k for $k = 1, \ldots, L$ to have mean=0, variance=1 for each stock symbol, $i = 1, \ldots, N$.
3. Set up a rectangular matrix G in Equation (II-2) and construct a correlation matrix C in Equation (II-7).
4. Diagonalize C to obtain N eigenvalues and discard eigenvalues smaller than λ_+.
5. The eigenvalues selected in 4 are sorted in descending order, such as $\lambda_1 > \lambda_2 > \cdots > \lambda_n > \lambda_+$.
6. Identify the industrial sectors of the dominant components of $\boldsymbol{v}_2$.

Fig. 3.1 Algorithm for extracting significant principal components in the RMT-PCA

$\boldsymbol{v}_{10}$. The data used for this is the 1-minute time series of TOPIX500 stock prices for the 3-year period 2007–2009 (Data-C in Table 2.2) and the results of tracking the time variation of each dominant sector using the principal components from $\boldsymbol{v}_2$ to $\boldsymbol{v}_{10}$. The results of tracking the time variation and the results of tracking the time variation of the dominant sector using only $\boldsymbol{v}_2$ in the analysis are shown and compared side by side.

We believe that tracking the dominant sectors using only $\boldsymbol{v}_2$ is a simpler and clearer representation of the changes in the industrial structure before and after the Lehman shock. In other cases, we used old data on US stock prices, daily data on the S&P 500 for 16 years from 1994 to 2009 (Data-B in Table 2.2), and analyzed the entire data as a single data set. The data are divided into two parts, one for each of the 8 years before and after and another part for each of the 4 years after the first part. The shorter the data, the more blurred the information becomes. The fact that no meaningful results were obtained when the data was further divided suggests that we have gained insight into the required data length.

RMT-PCA algorithm

1. (Preprocessing) $k = 1, \ldots, L + 1$, select N issue codes with at least one transaction at all time stamps. (Not necessary for regularly retrieved data)
2. Compute the log-return $r_i(t_k)$ for all N stocks selected. $k = 1, \ldots$. For each stock code $i = 1, \ldots, L$ time series t_k. L time series t_k. For each stock code $i = 1, \ldots, N$ so that mean $= 0$ and variance $= 1$ for each stock symbol $i = 1, \ldots, N$. 3.
3. Set up a rectangular matrix G by Eq. (2.2), and construct a correlation matrix C by Eq. (2.7).
4. Diagonalize C to obtain N eigenvalues and discard eigenvalues less than λ_+. Let the remaining eigenvalues be $\lambda_1, \lambda_+, \ldots,$ etc. in increasing order, and call the corresponding eigenvectors $\boldsymbol{v}_1, \boldsymbol{v}_2, \ldots$ (In the case of Japanese stocks, the first two digits of the four-digit stock code correspond approximately to the first two

digits of the four-digit stock code. For example, 95XX corresponds to electric power and gas supply companies, and 83XX to banks.)

We have collected and analyzed the tick data of the American market during the period of 1994–2002. The higher frequency of data points makes us possible to compare our results of different years and describe the historical change of the market in the scenario of this new method of principal component analysis based on RMT spectrum. For the sake of convenience, we named this methodology the RMT-PCA (Tanaka-Yamawaki et al. 2013). Here, tick data need a preprocessing of selecting the last price as the representative price in each block of 1-hour time interval. We call this as 'block-tick algorithm' and summarize it in Fig. 3.2. The reason to use $\boldsymbol{v}_2$ is as follows. The vector $\boldsymbol{v}_1$ has no dominant direction, nor $\boldsymbol{v}_3, \ldots, \boldsymbol{v}_n$; the only $\boldsymbol{v}_2$ carries information on the dominant directions.

In a previous study (Plerou et al. 2002), it was shown that the corresponding eigenvector components of the extracted principal components characterize prominent business sectors from 1991 to 1998. The previous study treated 7 years of data as a single set of data and was unable to track changes along the way. This is because the nature of RMT requires very long data. Consider the magnitude of N and L required for RMT, at the stage of estimating the distribution function from the N eigenvalues obtained by diagonalizing the correlation matrix and comparing it with the theoretical distribution. First, at least 100 eigenvalues are needed to infer the distribution function from the histogram of the eigenvalue distribution. In other words, $N > 100$ is required. To approximate a smoother function, N needs to be about 500. Then, the length of the time series, L, would need to be 1500 for the parameter value $Q = 3$. As long as daily data is used, the data length of 1 year is only about 220, so there would be no choice but to treat 7 years of data as one data set. Although this does not allow us to follow changes during the course of the year, it is sufficient for the verification of the method. In other words, with daily closing data, more than 7 years of data can ultimately be used for a single analysis. On the other hand, using intraday closing data would allow for one analysis per year, tracking year-to-year changes in active industry sectors.

We applied this approach to the various high-frequency data available at the time. However, intraday price data were not readily available at the time; prior to 2000, there were no instruments available to store large intraday stock prices. This situation rapidly improved around 2000. Not only did it become easier to

> The block-tick algorithm
>
> (1) Select N stock codes for which at least one transaction in all the 6 blocks of 10am=9:30-10:30, 3pm=2:30-3:30 on every working day of the year. Using the nearest data to the center of each block, such as 10am, etc., set up data for $k = 1, \ldots, L+1$, for N stocks.
>
> (2)-(6) are the same as the RMT-PCA in Fig. 3.1.

Fig. 3.2 The block-tick algorithm of RMT-PCA applied for the tick-wise stock prices

obtain large-capacity disks for storing data, but it also became possible to exchange data online through the Internet. At the same time, the Internet communication function in Windows 2000 enabled online stock trading. Inevitably, the volume of transactions increased and the intervals between transactions shortened. This became the era of big data accumulation, and from the perspective of RMT, extremely large N and L data was accumulated in all areas.

3.3 Annual Trends by Hourly Stock Price

Stock prices on the stock exchange appear to move randomly. For a long time, therefore, there was little interest in collecting tick data, and most records were either ignored or discarded. Recently, however, several institutions have begun to collect tick data. As a result, the statistical properties of price movements, such as the types of random walks and the mathematical properties of probability distributions, have become the subject of academic research. We studied high-frequency price movements of individual stocks and various stock indices from the JPX Cloud (JPX 2022) for the Japanese market and from the Refinitiv database (Refinitiv 2022) for other markets. The results show that stock prices are not simple random walks, nor are they independent (Tanaka-Yamawaki and Yamanaka 2019; Tanaka-Yamawaki and Ikura 2020, 2022). Certain stocks are strongly correlated with each other.

The resulting eigenvalues for the data of 1994 American stocks are plotted in Fig. 3.3 and compared to the theoretical distribution of the RMT-formula, using the parameters $N = 419$, $L = 1512$, $Q = L/N = 3.6$, to Eq. (2.28). First of all, the

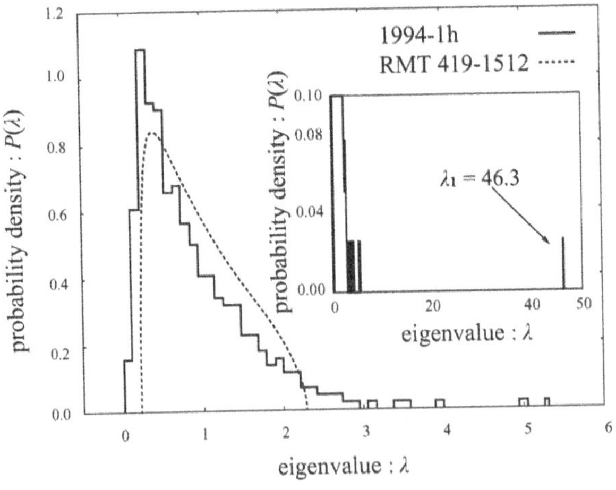

Fig. 3.3 Eigenvalues spectrum for 1994 NYSE ($N = 419$, $L = 1512$) and RMT in dotted line

Table 3.1 The values of N, L, $Q = L/N$, λ_+ in 1994, 1998, and 2002

Year	N	L	Q	λ_1	λ_2	λ_3	λ_4	...	λ_+
1994	419	1512	3.6	46	5.3	5.1	3.9	...	2.3
1998	490	1512	3.1	81	10	6.9	5.7	...	2.5
2002	569	1512	2.7	167	21	11	8.6	...	2.6

largest eigenvalue is $\lambda_1 = 46.3$, far larger than the theoretical maximum $\lambda_+ = 2.32$ of this case. There are ten more eigenvalues larger than λ_+. Clearly, the distribution deviates from Eq. (2.28) by those discrete eigenvalues,

$$\lambda_1 \gg \lambda_2 > \ldots > \lambda_+$$

On the other hand, the remaining smaller eigenvalues have a continuous spectrum close to the theoretical curve. Following the procedure described so far, we analyze data of 1994–2002, year by year, and obtain the distribution of eigenvalues. The largest few eigenvalues are the focus of our attention. In Table 3.1, the largest four eigenvalues are tabulated, together with parameters, N, L, $Q = L/N$, λ_+, for 1994, 1998, and 2002.

Looking at Fig. 3.3, for example, the eigenvalue distribution (histogram) and the corresponding RMT theoretical curve (dotted line) roughly overlap in the range of $\lambda < \lambda_+$, and several large eigenvalues $\lambda_+ < \lambda_2 \ll \lambda_1$ appear outside of the range of RMT. This is how RMT-PCA separates the principal components from the background noise. The corresponding eigenvector components for the noise part are highly random, while the eigenvector components for the principal components show clear signals of aggregation.

The eigenvector $\boldsymbol{v}_1$ corresponding to the largest eigenvalue λ_1 is the first principal component. For 1-hour data of 1994, the major components of $\boldsymbol{v}_1$ are giant companies such as GM, Chrysler, JP Morgan, Merrill Lynch, and DOW Chemical. Although the first principal component is said to be the market mode, our result in this analysis simply shows that the giant companies have major contributions to the eigenvector components. In 1998 and 2002, the major components of $\boldsymbol{v}_1$ are dominated by both financial companies. Since the major business sector moved from manufacturers to financial companies around the change of the century, this is a convincing result.

The eigenvector $\boldsymbol{v}_2$ corresponding to the second largest eigenvalue λ_2 is the second principal component. For 1-hour data of 1994, the major components of $\boldsymbol{v}_2$ are dominated by seven mining companies and two finance companies. For 1998, the dominant part becomes ten electric companies and then in 2002 becomes six food companies.

The eigenvector $\boldsymbol{v}_3$ corresponding to the third largest eigenvalue λ_3 is the third principal component. For 1-hour data of 1994, the major components of $\boldsymbol{v}_3$ consist of semiconductor manufacturers, including Intel. For 1998, this is changed to banks and financial services and then in 2002 becomes six food companies. The results

Table 3.2 The sectors extracted from v_1, v_2, and v_3 for 1994, 1998, and 2002 compared to 1990–1996

1990–1996	v_1	(Diverse sectors)
(Plerou et al. 2002)	v_2	Industrials (6) + InfoTech (4)
	v_3	Materials (8) + Industrials (1) + InfoTech (1)
1994	v_1	Finance (4) + InfoTech (2) + Industrials (3)
	v_2	Materials (7) + Finance (2)
	v_3	InfoTech (10)
1998	v_1	Finance (8)
	v_2	Utility (10)
	v_3	Finance (3)
2002	v_1	Finance (9)
	v_2	Foods (6)
	v_3	Utility (10)

are summarized in Table 3.2. Here, the tick data requires preprocessing to select the last price as the representative price in each block of 1-hour time intervals. This block-tick algorithm is outlined in Fig. 3.2.

In this first trial, we attempted to check the methodology by using the block-tick method to extract the equal-time data in the $N \times L$ rectangular shape, to construct the equal-time correlation matrix of N stock time series. Although the result failed to reproduce the market mode for the elements of the first eigenvector v_1, the eigenvalue distribution in Fig. 3.3 and Table 3.1 and the major industrial sectors of v_2 and v_3 in Table 3.2 exhibit reasonable results.

3.4 Annual Trends of Major Sectors on NYSE

The results of the first lesson were somewhat disappointing, perhaps because the tick data from 1994 to 2002 were not dense enough for the block-tick method to work well. As a result of picking up a very small number of representative prices from a less-dense price record, the formatted data may have suffered from noise of various kinds. In fact, the size of the database has grown by several orders of magnitude over the past 20 years. Therefore, in a second attempt, we decided to analyze the NYSE daily data for a longer period of 16 years, from 1994 to 2009. As shown in Table 3.3, sufficiently active $N = 373$ stocks with data length $L = 3961$, for the period of 1994–2009, are prepared. By dividing this data into two periods, a data for the period of 1994–2001 with $L = 2015$ and another data for the period of 2002–2009 with $L = 1946$ are prepared and analyzed by RMT-PCA. We have tried further to split to four periods of 1994–1997, 1998–2001, 2002–2005, and 2006–2009, and the resulting parameters N, L, $Q = L/N$, and λ_+ are summarized in Table 3.4. Further splitting the data into eight parts of 2 years each did not yield

Table 3.3 Results for 16-, 8-, and 4-year NYSE market daily data

	Year	N	L	Q	λ_1	λ_2	λ_3	λ_4	$\cdots$	λ_+
16 yrs.	1994–2009	373	3961	10.6	74	11	8.8	7.7	$\cdots$	1.7
8 yrs.×2	1994–2001	373	2015	5.40	41	13	8.8	6.9	$\cdots$	2.1
	2002–2009	464	1946	4.19	150	15	12	11	$\cdots$	2.2
4 yrs.×4	1994–1997	373	1010	2.71	37	8.7	5.8	4.6	$\cdots$	2.6
	1998–2001	419	1002	2.39	53	19	13	9.2	$\cdots$	2.8
	2002–2005	464	1006	2.17	116	14	13	9.1	$\cdots$	2.8
	2006–2009	468	936	2.00	200	18	14	8.9	$\cdots$	2.9

Table 3.4 The list of GICS code to classify the business sectors of stocks is shown

A: Energy	F: Healthcare product
B: Material	G: Finance
C: Industrial	H: Information technology
D: Service	I: Telecommunications
E: Consumer Products	J: Utility

meaningful results. This experience gave us a hint of the minimum length L needed for this method.

The data length from 1994 to 2009 is $L = 3961$. Among the $N = 373$ firms corresponding to the dimensions of the eigenvector, we identify the industries of the 20 largest components of the corresponding eigenvector. If these components are concentrated in a particular industry, identify that industry as the trend maker, or the major industrial sector, for that period.

The sector classification follows the Global Industry Classification Standard (GICS), a code system that classifies stock industries into ten categories. As shown in Table 3.4, the codes are represented by a single uppercase letter from A to J.

Unlike the first attempt to extract the isochronous N-by-L rectangular matrix from the tick data using the block-tick method, this time we expect better results and use the daily prices of $L = 3961$ stocks of $N = 373$ in the isochronous N-by-L rectangular matrix for 16 years as given data.

The largest principal component is given by the largest eigenvalue λ_1 and its eigenvector $\boldsymbol{v}_1$. This is not suitable for the leading sector: the magnitude of the N components of $\boldsymbol{v}_1$ is approximately equal to the mean $1/\sqrt{N}$, with no outstanding component. This fact is common to most markets and is often referred to as the "market mode." This component is known to be strongly correlated with indices composed of dominant and stable stocks, such as so-called blue-chip stocks.

The second principal component, eigenvector $\boldsymbol{v}_2$ for the second highest eigenvalue λ_2, has a wavy structure and shows coherent movements of adjacent companies belonging to the same industry. Therefore, we focus on the second and upper principal components to pick out the leading business sectors in each period.

The results for the entire 16-year period are shown in Fig. 3.4. Each bar shows the percentage of industry sectors to which the 20 largest positive and negative components belong. The eight bars are aligned from left to right, with the corresponding

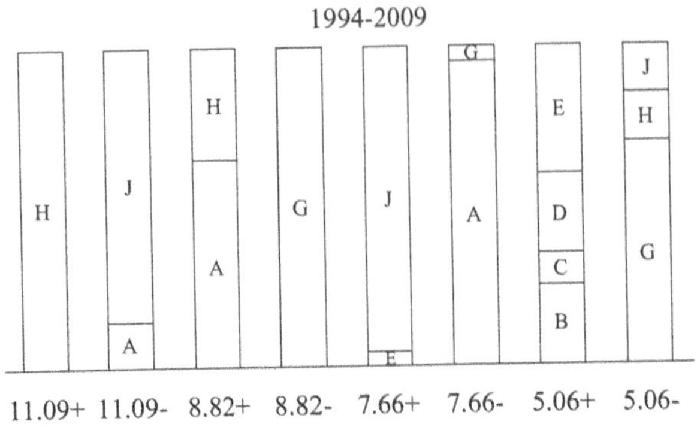

Fig. 3.4 The leading sectors for 16 years (1994–2009) are H (IT) and J (Utility) shown at 11.09

eigenvalues shown below the bars. For example, the two leftmost bars represent the (+) and (−) components of $\boldsymbol{v}_2$. The numbers below the bars indicate the corresponding eigenvalues $\lambda_2 = 11.09$. The (+)/(−) denotes the positive/negative components of the eigenvalue $\boldsymbol{v}_2$. In this case, the 20 largest positive components are all "H" representing the "Information technology" sector, the 20 largest negative components are mostly "J" representing the "Public utilities" sector with more than 80%, but a small portion of the 20 or less is "A" representing the "Energy" sector. Similarly, the third and fourth bars are the (+) and (−) components of $\boldsymbol{v}_3$. The numbers below the bars indicate the corresponding eigenvalue $\lambda_3 = 8.82$. The positive component of $\boldsymbol{v}_3(+)$ is more A (Energy) and less H (InfoTech), while the negative component of $\boldsymbol{v}_3(-)$ is more G (Finance).

Next, we split the total 16-year period into two 8-year periods, 1994–2001 and 2002–2009, to check consistency. The second principal component $\boldsymbol{v}_2$ for 1994–2001 in the left graph is dominated by H (Information technology) and J (Utilities) as in the 1994–2009 graph. On the other hand, the second principal component $\boldsymbol{v}_2$ in the right graph is dominated by A (energy) and G (financial) in 2002–2009. The results for the entire 16-year period are shown in Fig. 3.5.

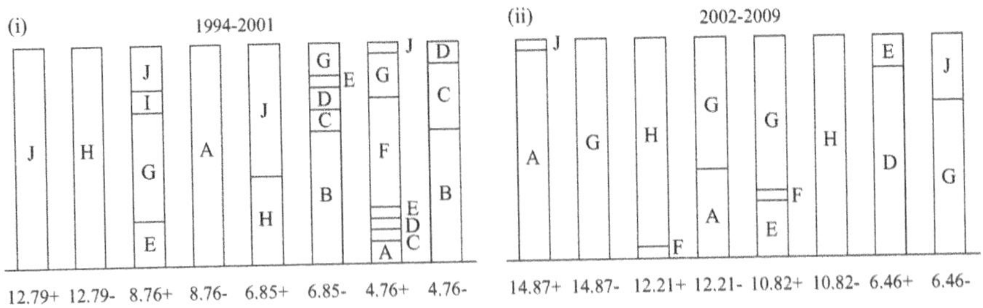

Fig. 3.5 The leading sectors change from J and H (1994–2001) at 12.79 to A and G (2002–2009) at 14.87

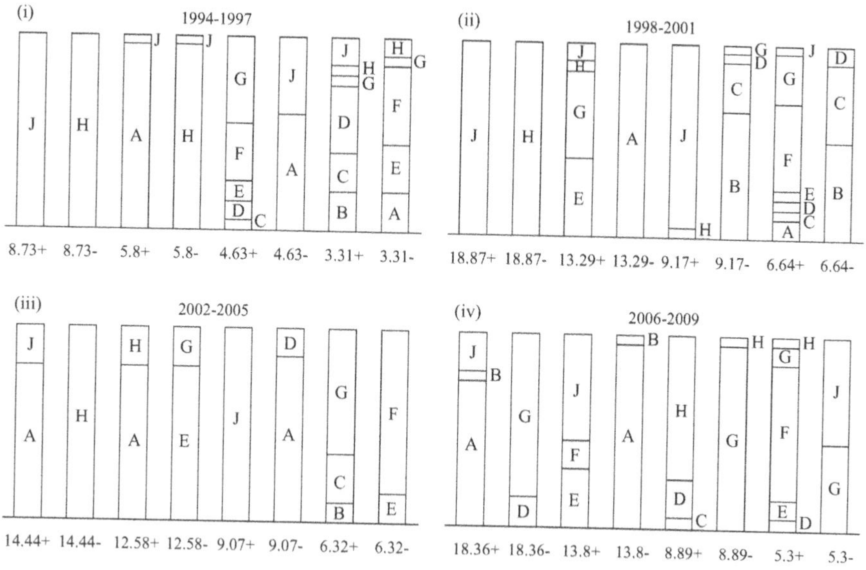

Fig. 3.6 The leading sectors change from J and H (94-01) to A and H (02-05) and then to A and G (06-09)

What if we further divide the data into four parts? As shown in Fig. 3.6, $\boldsymbol{v}_2$ in 1994–1997 and 1998–2001 in the above figure is dominated by J (Utility) and H (Info tech), which is the same feature as in 1994–2001. $\boldsymbol{v}_2$ in 2002–2005 is dominated by A (Energy) and H (Info tech), which is very different from the 2002–2009 case. However, $\boldsymbol{v}_2$ for 2006–2009 is almost the same as in the 2002–2009 case. Thus, the positive and negative components of the eigenvectors corresponding to the second principal components show that the trend in each period is well represented. $\boldsymbol{v}_2$ results are shown in Table 3.5 (Tanaka-Yamawaki et al. 2013).

3.5 Quarterly Trends of Tokyo Market

The data used in this section is 1-minute time series for the 500 stocks listed in the TOPIX500 list in 2007. The data period was 3 years from 2007 to 2009. Considering that the 2 years was the shortest analysis period for the daily closing prices available on the web, due to the limitations of the number of data and data length of the RMT-PCA, which was discussed in Sect. 3.4, by conducting the analysis using daily-close price, it became possible to follow changes over a year period, successfully shortening the analysis period. In this study, we further shortened the analysis period to every 3 months. We also attempted to track changes in major industry sectors on a monthly basis by using data from overlapping periods, such as January to March and February to April as shown in Table 3.6.

Table 3.5 The leading sectors appearing in v_2 are tabulated

1994–2009	$v_2(+)$	InfoTech: H (20 companies)
	$v_2(-)$	Utility: J (17 companies) + Energy: A (3 companies)
1994–2001	$v_2(+)$	Utility: J (20)
	$v_2(-)$	InfoTech: H (20)
2002–2009	$v_2(+)$	Energy: A (19) + Utility: J (1)
	$v_2(-)$	Finance: G (20)
1994–1997	$v_2(+)$	Utility: J (20)
	$v_2(-)$	InfoTech: H (20)
1998–2001	$v_2(+)$	Utility: J (20)
	$v_2(-)$	InfoTech: H (20)
2002–2005	$v_2(+)$	Energy: A (16) + Utility: J (4)
	$v_2(-)$	InfoTech: H (20)
2006–2009	$v_2(+)$	Energy: A (15) + Utility: J (4) + Material: B (1)
	$v_2(-)$	Finance: G (17) + Service: D (3)

Table 3.6 The values of N, L, Q, λ_1, λ_2, λ_3, and λ_+ for the 12 quarters in 2007–2009

Year	N	L	Q	λ_1	λ_2	λ_3	$\cdots$	λ_+
2007	477	2682	5.62	118	22.9	10.4	$\cdots$	2.02
2008	480	2682	5.59	170	13.0	7.60	$\cdots$	2.03
2009	476	2655	5.58	149	18.8	8.60	$\cdots$	2.03
2007-I	486	642	1.32	139	33.4	11.7	$\cdots$	3.50
2007-II	486	681	1.40	101	15.0	10.2	$\cdots$	3.40
2007-III	489	681	1.39	130	24.9	14.1	$\cdots$	3.41
2007-IV	492	675	1.37	139	16.9	8.86	$\cdots$	3.44
2008-I	488	642	1.32	189	14.5	8.15	$\cdots$	4.20
2008-II	491	681	1.39	148	10.9	8.84	$\cdots$	4.10
2008-III	492	692	1.41	155	14.5	13.5	$\cdots$	4.08
2008-IV	487	664	1.36	197	19.2	14.9	$\cdots$	4.14
2009-I	490	642	1.31	157	20.3	9.00	$\cdots$	3.51
2009-II	486	659	1.36	149	17.6	9.20	$\cdots$	3.46
2009-III	485	681	1.40	145	18.1	6.99	$\cdots$	3.40
2009-IV	483	670	1.39	121	20.1	9.27	$\cdots$	3.42

The Tokyo Security Exchange (TSE) which has been converted to Japan Exchange (JPX) has a convenient numbering system for stock coding. That is, stocks are numbered in order of industry sector, with some exceptions, such as various financial indices, construction, and fisheries in the 1000s, chemicals and pharmaceuticals in 4000s, resources and materials in the 5000s, machineries and electronics in the 6000s, banking and finance in the 8300s, and electric/gas power suppliers in the 9500s, and stocks included in TOPIX500 follow this numbering system. The Stocks in the TOPIX 500 are basically classified by industry based on the first two digits of the code number. The code numbers of the 483 stocks whose

Table 3.7 The industrial sectors represented by the stock code numbers in Tokyo Market

*	Sectors	Stock code	Stock code actually included	**
13	Fisheries, Agriculture, Forestry	1300-1499	1332	1
15	Mining	1500-1699	1605,1662	16
17	Construction	1700-1999	1721,1801,1802,1803,1808,1812,1820,1833,1860,1878,1911,1925,1928,1944, 1951,1963	16
20	Foodstuffs	2000-2999	2002,2202,2212,2267,2282,2331,2432,2433,2501,2502,2503,2531,2579,2593, 2651,2685,2730,2273,2768,2784,2801,2809,2810,2811,2871,2875,2897,2914	28
30	Textiles and paper	3000-4999	3002,3086,3088,3101,3105,3116,3231,3382,34013402,3404,3405,3407,3436, 3591,3861,3880,3893,3941	19
40	Chemicals (-4499), Pharmaceuticals service (-4599), others (4600-)	4000-4999	4004,4005,4021,4042,4043,4044,4045,4061,4062,4063,4088,4091,4114,4118, 4151,4182,4183,4185,4186,4188,4202,4203,4204,4205,4208,4217,4272,4307, 4324,4401,4452,4502,4503,4506,4507,4508,4519,4521,4523,4527,4528,4530, 4534,4535,4536,4540,4543,4544,4547,4568,4612,4613,4631,4661,4665,4666, 4676,4684,4689,4704,4716,4732,4739,4768,4901,4902,4911,4912,4922,4967	70
50	Resources, Materials	5000-5999	5001,5002,5007,5012,5016,5019,5101,5108,5110,5201,5202,5214,5232,5233, 5301,5332,5333,5334,5401,5405,5406,5407,5411,5423,5444,5451,5463,5471, 5481,5482,5486,5541,5631,5701,5706,5711,5713,5714,5726,5727,5801,5802, 5803,5812,5855,5901,5929,5938,5947,5991	50
60	Machinery, Electric appliances	6000-6999	6103,6113,6136,6141,6146,6201,6222,6268,6273,6301,6302,6305,6326,6349, 6361,6366,6367,6370,6383,6395,6417,6448,6457,6460,6471,6472,6473,6474, 6479,6481,6501,6502,6503,6504,6506,6581,6586,6588,6592,6594,6645,6665, 6674,6701,6702,6703.6707,6723,6724,6728,6752,6753,6758,6762,6764,6767, 6770,6773,6804,6806,6841,6845,6849,6856,6857,6861,6869,6902,6923,6925, 6952,6954,6963,6965,6967,6971,6976,6981,6986,6988,6991,6995,6996	83
70	Automobiles, Transportation equipment	7000-7999	7003,7004,7011,7012,7013,7201,7202,7203,7205,7211,7221,7230,7240,7241, 7251,7259,7261,7262,7267,7269,7270,7272,7276,7282,7309,7312,7453,7459, 7518,7532,7649,7701,7729,7731,7733,7735,7741,7751,7752,7762,7832,7911, 7912,7915,7936,7951,7966,7974,7984,7988	50
80	Commercial	8000-8299	8001,8002,8012,8015,8016,8028,8031,8035,8036,8053,8056,8058,8060,8078, 8086,8112,8113,8129,8184,8218,8219,8227,8233,8242,8252,8253,8267,8270, 8273,8282	30
83	Finance, Insurances	8300-8799	8303,8304,8306,8308,8309,8316,8324,8327,8328,8331,8332,8333,8334,8336, 8339,8341,8354,8355,8356,8358,8359,8361,8363,8366,8368,8369,8377,8379, 8381,8382,8385,8386,8388,8390,8394,8403,8404,8411,8415,8418,8473,8511, 8515,8522,8544,8564,8570,8572,8574,8586,8591,8593,8595,8601,8604,8606, 8607,8609,8616,8628,8698,8703,8729,8754,8755,8759,8761,8763,8766,8795	70
88	Real estate	8800-8999	8801,8802,8804,8815,8830,8848,8905,8933	8
90	Transportation, Telecommunication	9000-9499	9001,9005,9006,9007,9008,9009,9020,9021,9022,9031,9041,9042,9045,9048, 9062,9064,9065,9076,9101,9104,9107,9132,9202,9205,9301,9303,9364,9401, 9404,9409,9432,9433,9435,9437	30
95	Electric/Gass power supply	9500-9599	9501,9502,9503,9504,9505,9506,9507,9508,9509,9513,9531,9532, 9533	13
96	Communication, Retail, Service	9600-9999	9602,9613,9684,9697,9706,9735,9737,9744,9747,9766,9783,9793,9831,9832, 9843,9861,9962,9983,9984,9987,9989	21

prices are treated in this chapter are summarized according to this classification in Table 3.7. Note that in the first column of this table depicted as (∗), the symbols used in the following figures are shown. For example, the symbol (17) is used to represent the code numbers from 1700 to 1999.

However, as shown in the right most column of Table 3.7 depicted as (∗∗), the number of listed stocks is not evenly distributed across industries. For example, only one issue 1332 is chosen for the category (13), and two for (15), while 83 issues are chosen for the category (60). Moreover, stocks in industries that have grown rapidly in recent years, such as information and telecommunications, do not have enough code numbers in the 9000 range alone, but are assigned to the 2000–4000 range, making it difficult to find industry bias. Still, the result shown in the following figures is not strongly affected by this kind of mis-classification in the coding. This is mainly because the major industrial sectors shifted from the category (83) to the category (95) in the period of 2007–2009.

The advantage of this industry-based code numbering system is apparent in the graphs of eigenvector components of the principal components. When the N components are plotted according to the code numbers, just because the companies in the type of industry are assigned the same two-digit number, e.g., 7201, 7202, 7203, and 7205, are all automobile manufacturers, 9501–9509 are electric power companies, 8300 numbers are banks, the N-dimensional eigenvector components are automatically sorted in ascending order by industry sector.

Aided by this code numbering system, graphing the N-dimensional eigenvector components makes it easy to instantly recognize the location of strong and weak sectors in the case of TSE stock prices.

For example, as a typical solution to the eigenvalue problem

$$C\boldsymbol{v}_k = \lambda_k \boldsymbol{v}_k \quad (k = 1, \ldots, N)$$

for the correlation matrix C for the stock prices of TSE or, equivalently, diagonalizing C to obtain the eigenvalues λ_k and the eigenvectors $\boldsymbol{v}_k$, where k is renumbered in the descending order of λ as in the algorithm of Fig. 3.1.

We show $k = 1, \ldots, 4$ eigenvector components $\boldsymbol{v}_k$ and their probability distributions in Fig. 3.7, from which we can see that the N components of eigenvector $\boldsymbol{v}_1$ corresponding to the largest eigenvalue λ_1 are flatly distributed, as shown in the left panel of the first row of Fig. 3.7. The components are all positive and have almost the same magnitude of 0.05. Considering that the magnitude of each component is about $1/\sqrt{500}$, which is about $1/22$, this distribution is considered to be consistent. The right panel of the first row of Fig. 3.7 shows the probability density distribution of the components, which is strongly biased toward the positive side and peaked at around 0.05.

It has been mathematically proven (Perron 1907; Frobenius 1912) that the eigenvector corresponding to the largest eigenvalue of a positive matrix consists of components of the same sign, known as the Perron-Frobenius theorem. It is also known that the components of the eigenvector corresponding to the largest eigenvalue of the correlation matrix of stock prices in the stock market are not concentrated in a particular sector and are evenly distributed. For this reason, the first principal component is not suitable for extracting information on strong sectors. On the other hand, the eigenvectors of the other eigenvalues have components with positive and negative signs. Therefore, for the purpose of finding strong sectors, the second principal component is the next candidate of interest.

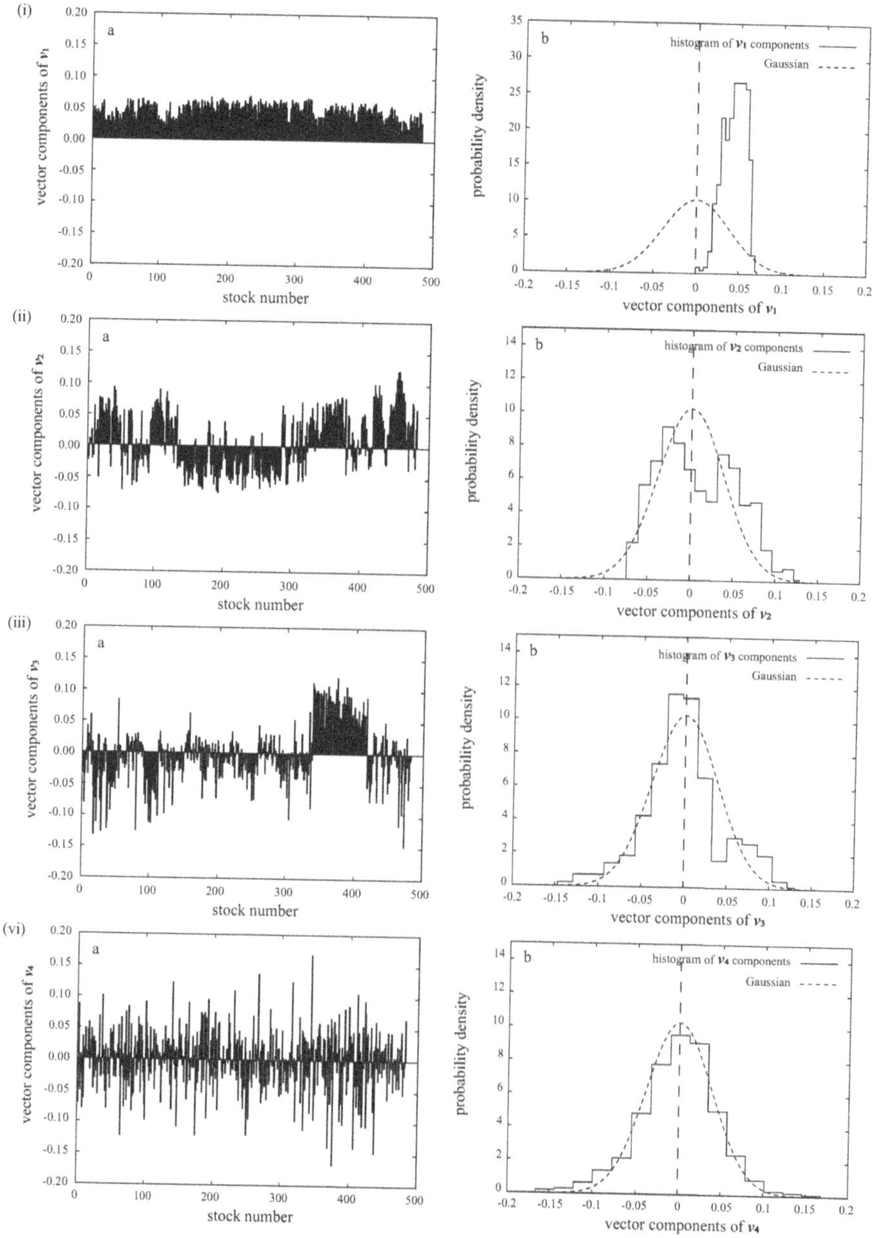

Fig. 3.7 Components (i, x_i) of the first,...,fourth eigenvectors $\boldsymbol{v}_1, \ldots, \boldsymbol{v}_4$ are plotted for $i = 1, \ldots, N$ (left) and the histogram in solid line, compared to the normal distribution in dotted line (right)

The second principal component $\boldsymbol{v}_2$ on the left panel of the second row has a wavy structure. The highest peak is observed near the right end, indicating that the public service sector is developed in the range of code number 9000. However, the left side of the third principal component $\boldsymbol{v}_3$ gives less clear signal than the $\boldsymbol{v}_2$ component. The fourth principal component $\boldsymbol{v}_4$ is even worse and is considered random noise. The probability distribution shown on its right side is approximately equal to the Gaussian distribution shown by the dotted line. Thus, the most useful component appears to be the second eigenvector $\boldsymbol{v}_2$. As shown in Fig. 3.7, the peak of the component of $\boldsymbol{v}_2$ appears with its companion and moves in waves. Thus, it can be seen that the component of $\boldsymbol{v}_2$ acts as a coherent mode constituting the leading sector of the moment.

The result of applying RMT-PCA on this 1-minute price data for 483 stocks listed in TOPIX500 in 2007–2009 is summarized in Table 3.6. For each year and each quarter, leading sectors were extracted from up to five components, and corresponding industry sectors were identified.

After separating the principal components from the random noise by using RMT-PCA, a question remains on how many principal components play important roles. Although there are more than ten eigenvalues that are larger than the RMT limit λ_+, many of them are close to the limit. Moreover, correlation between prices of different stocks is apparent in the components of $\boldsymbol{v}_2$, the second eigenvector. In comparison to this, components of the third eigenvector $\boldsymbol{v}_3$ appear to be less correlated, as shown by Fig. 3.7, and the correlation practically diminishes for fourth or higher principal components.

As an intermediate process, we determine to use ten components, corresponding to the largest five eigenvalues, and picked up 20 prominently large (in magnitude) components and identified the corresponding industrial sectors. The results are shown in Fig. 3.8 for each 6-month period in 2007–2009. The magnitude of partitions is determined by the occupation rate of individual sector. For example, if only the first two components $v_{k,1}$ and $v_{k,2}$ in an eigenvector

$$\boldsymbol{v}_k = \begin{pmatrix} v_{k,1} \\ v_{k,2} \\ \vdots \\ v_{k,N} \end{pmatrix}$$

belong to sector "83" and these two components are the selected members of 20 large components, then the magnitude of the partition for "83" is computed by fraction of the squared sum of those components belonging to sector "83" over the total sum of 20 selected components, such as

$$\left(v_{k,1}^2 + v_{k,2}^2\right) \Big/ \sum_{m=1}^{20} v_{k,m}^2$$

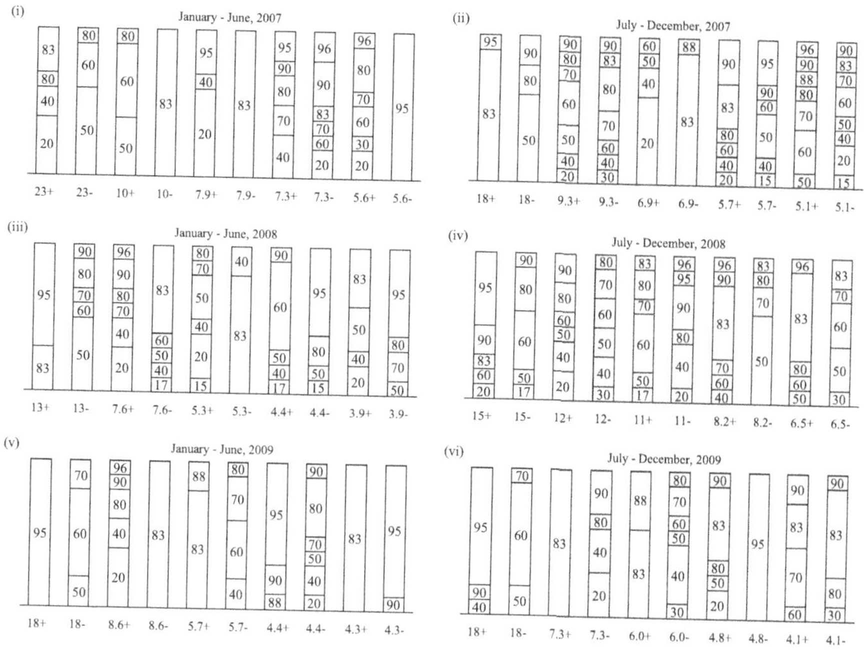

Fig. 3.8 Quarterly change of major industrial sectors in 2007 by means of ten principal components

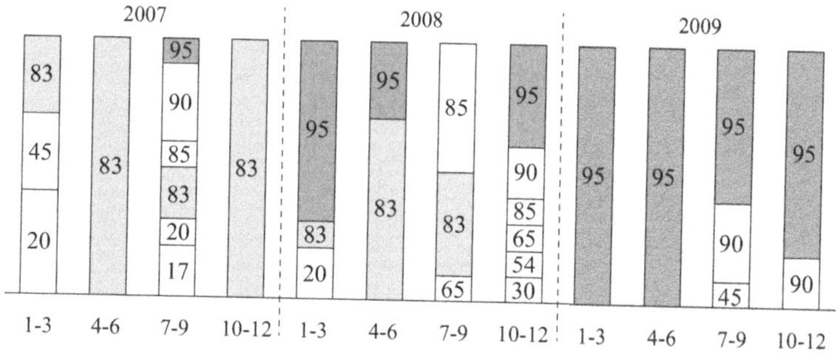

Fig. 3.9 Quarterly change of major sectors extracted in 2007–2009 by using from $\boldsymbol{v}_2$

However, of relative importance the eight bars in each panel of Fig. 3.8 are not equal. We can also draw graphs simply based on the second eigenvector $\boldsymbol{v}_2$. This is shown in Fig. 3.9.

We discuss whether these results are consistent with actual historical incidence. Comparing the annual and quarterly graphs in Fig. 3.8, there is a gradual shift in the leading industry sector from "83" (banking) in 2007 to "95" (electricity supply) in 2009. In addition, the disappearance of major sectors in the third quarter of 2007 and the fourth quarter of 2008 indicates the extreme market turmoil caused by the subprime mortgage crisis in August 2007 and the Lehman shock in October 2008.

From this vast amount of price fluctuation data, we have succeeded in extracting changes in stock market trends. In particular, as shown in Fig. 3.9, we succeeded in detecting the historical breakdown of the subprime loan problem in the third quarter of 2007 and the Lehman shock in the fourth quarter of 2008.

Chapter 4
The RMT-Tests

4.1 Motivation

We created RMT-test as a new tool to measure the randomness of sufficiently long data strings and present the results of applying it to various situations. The goal of RMT-test is to find good labels for a given data string. Good labels can greatly reduce the burden of big data analysis, especially for huge financial data.

RMT-test quantifies the degree of correlation between pairs of different parts of a data string, thereby measuring the randomness of a single data string. Unlike the RMT-PCA in Chap. 2, the matrix elements of the correlation matrix C are not inner products between the price vectors of different stocks, but between different parts taken from a single data string.

Once the correlation matrix C is determined, the subsequent process of computing the eigenvalues is technically identical to the RMT-PCA described in the first half of this book, except that RMT-test uses the distribution of eigenvalues to calculate randomness, whereas RMT-PCA analyzed the components of eigenvectors corresponding to the principal components. Therefore, the first half of Chap. 4, especially the Sect. 4.2, has a lot of overlap with Chap. 2. However, referring to Chap. 2 in the middle of Chap. 4 for fear of duplication is not only cumbersome but also confusing, so the description here is written to be completed only in Chap. 4 without worrying about duplication.

There are virtually no restrictions on the selection of this data string, except that it must be long enough and have at least 30,000 data points to use $N = 100$ and 120,000 data points to use $N = 200$. On the other hand, the data used in the RMT-PCA in Chaps. 2 and 3 are more restrictive. The data structure is an N-by-L rectangular matrix, which is required to have equal time data for each of N different targets, e.g., N different stock price time series. Actually, this is quite difficult to satisfy. Although daily-close price data automatically satisfy simultaneity, a very long period of time is required to satisfy the condition $N > 500$, say, some

M. Tanaka-Yamawaki, Y. Ikura, *Principal Component Analysis and Randomness Test for Big Data Analysis*, Evolutionary Economics and Social Complexity Science 25, https://doi.org/10.1007/978-981-19-3967-9_4

years. Intra-day price such as tick-wise prices, on the other hand, has a problem of simultaneity, since different stock prices are traded at different times; a tedious preprocessing is required to set up data for the rectangle of $N \times L$. In short, RMT-PCA has limited applicability. The RMT-test has no such difficulty and is expected to have much wider applicability than RMT-PCA. After introducing a naive qualitative version, we proceed to a quantitative version of the RMT-test.

In a qualitative version, the eigenvalue spectrum of the autocorrelation matrix is simply compared to its theoretical counterpart derived by RMT. This method is effective for data with relatively low levels of randomness, since the differences can be easily recognized by looking at the figure. On the other hand, when the randomness is very high, the eigenvalue spectrum of random data and the RMT curve become indistinguishable. Therefore, we developed a quantitative version of the RMT-test and introduced two methodologies to quantify the degree of randomness in the later sections.

The first measure of randomness is the deviation of the k-th moment from the theoretical value of the RMT, D_k. The second measure is the distance $\Delta\lambda = \lambda_1 - \lambda_+$ between the greatest eigenvalue λ_1 and the RMT theoretical upper bound λ_+, which is mainly used for the case of relatively low randomness.

4.2 Formulation: Basic Formulas

Consider measuring the randomness of a very long one-dimensional array. Cut it into L intercepts of equal length and number them by $i = 1, \ldots, N$. Discarding the remainder of length $L - 1$, we obtain the following N vectors of dimension L

$$\boldsymbol{g}_i = (g_{i,1} \; g_{i,2} \; \cdots \; g_{i,L}) \tag{4.1}$$

where i is the position of the block. Here as in Sect. 2.1, we normalize the sequence of data

$$g_{i,k} = \frac{r_{i,k} - \langle r_i \rangle}{\sigma_i} \quad (i = 1, \ldots, N, \; k = 1, \ldots, L) \tag{4.2}$$

and

$$\langle r_i \rangle = \frac{1}{L} \sum_{k=1}^{L} r_{i,k} \tag{4.3}$$

and

$$\sigma_i^2 = \langle r_i^2 \rangle - \langle r_i \rangle^2 \tag{4.4}$$

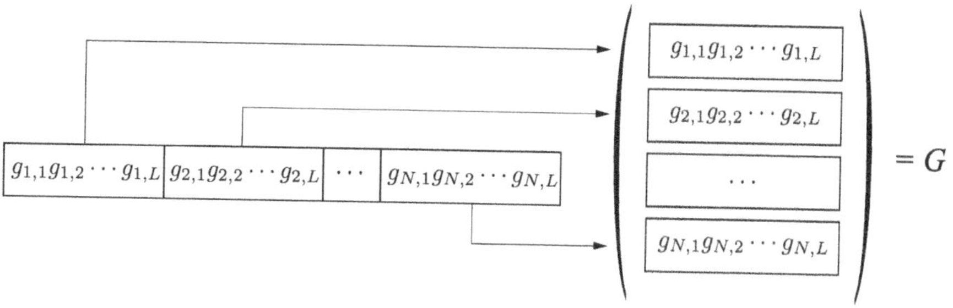

Fig. 4.1 Making the data for the RMT-test from the original data

Unlike the RMT-PCA case examined in Chaps. 2–3, the suffix $i = 1, \ldots, N$ indicates the position of $\boldsymbol{g}_i$ in the original data sequence. In other words, as shown in Fig. 4.1, the original data sequence $g_{1,1}, g_{1,2}, \ldots, g_{N,L}$ can be restored by concatenating $\boldsymbol{g}_i$.

Next, an N-by-L rectangular matrix is constructed as follows.

$$G = \begin{pmatrix} \boldsymbol{g}_1 \\ \boldsymbol{g}_2 \\ \vdots \\ \boldsymbol{g}_N \end{pmatrix} = \begin{pmatrix} g_{1,1} & g_{1,2} & \cdots & g_{1,L} \\ g_{2,1} & g_{2,2} & \cdots & g_{2,L} \\ \vdots & \vdots & \ddots & \vdots \\ g_{N,1} & g_{N,2} & \cdots & g_{N,L} \end{pmatrix} \tag{4.5}$$

Its transverse G^t is an L-by-N rectangular matrix as follows.

$$G^t = \begin{pmatrix} g_{1,1} & g_{2,1} & \cdots & g_{N,1} \\ g_{1,2} & g_{2,2} & \cdots & g_{N,2} \\ \vdots & \vdots & \ddots & \vdots \\ g_{1,L} & g_{2,L} & \cdots & g_{N,L} \end{pmatrix} \tag{4.6}$$

The autocorrelation matrix C is constructed by multiplying the rectangular matrix G and its transverse G^t as follows.

$$C = GG^t = \begin{pmatrix} c_{1,1} & c_{1,2} & \cdots & c_{1,N} \\ c_{2,1} & c_{2,2} & \cdots & c_{2,N} \\ \vdots & \vdots & \ddots & \vdots \\ c_{N,1} & c_{N,2} & \cdots & c_{N,N} \end{pmatrix} = \begin{pmatrix} \boldsymbol{g}_1 \cdot \boldsymbol{g}_1 & \boldsymbol{g}_1 \cdot \boldsymbol{g}_2 & \cdots & \boldsymbol{g}_1 \cdot \boldsymbol{g}_N \\ \boldsymbol{g}_2 \cdot \boldsymbol{g}_1 & \boldsymbol{g}_2 \cdot \boldsymbol{g}_2 & \cdots & \boldsymbol{g}_2 \cdot \boldsymbol{g}_N \\ \vdots & \vdots & \ddots & \vdots \\ \boldsymbol{g}_N \cdot \boldsymbol{g}_1 & \boldsymbol{g}_N \cdot \boldsymbol{g}_2 & \cdots & \boldsymbol{g}_N \cdot \boldsymbol{g}_N \end{pmatrix} \tag{4.7}$$

where each element of the correlation matrix C at the i-th row and j-th column is the inner product of the i-th vector, $\boldsymbol{g}_i$, and the j-th vector, $\boldsymbol{g}_j$, defined as follows.

$$\boldsymbol{g}_i \cdot \boldsymbol{g}_j = \sum_{k=1}^{L} g_{i,k} g_{j,k} \tag{4.8}$$

The elements of the i-th row and the j-th column of matrix C are expressed as the (i, j) components of C and are written as $C_{i,j}$. By the definition of Eq. (4.7), C is symmetric about the exchange of i and j, and we have

$$C_{i,j} = C_{j,i} \quad (i = 1, \ldots, N; \; j = 1, \ldots, N) \tag{4.9}$$

Also, since all diagonal elements are equal to 1 by the normalization condition of Eq. (2.5)

$$C_{i,i} = 1 \quad (i = 1, \ldots, N) \tag{4.10}$$

Furthermore, the absolute value of its off-diagonal elements is generally less than 1

$$|C_{i,j}| \leq 1 \quad (i = 1, \ldots, N; \; j = 1, \ldots, N) \tag{4.11}$$

Recall that the inner product of two vectors $\boldsymbol{g}_i$ and $\boldsymbol{g}_j$ is defined as $\boldsymbol{g}_i \cdot \boldsymbol{g}_j = |\boldsymbol{g}_i||\boldsymbol{g}_j| \cos\theta$, where θ is the angle between $\boldsymbol{g}_i$ and $\boldsymbol{g}_j$. In the case of unit vectors, $|\boldsymbol{g}_i| = |\boldsymbol{g}_j| = 1$, the magnitude of the inner product is $|\cos\theta| \leq 1$. From this definition, each element $C_{i,j}$ of the correlation matrix C represents the degree of overlap between the two price time series vectors $\boldsymbol{g}_i$ and $\boldsymbol{g}_j$.

As is well known, any real symmetric matrix C can be diagonalized by the similarity transformation $V^{-1}CV = V^tCV$ by an orthogonal matrix V satisfying $V^{-1} = V^t$, where V^{-1} and V^t represent the inverse matrix and the transverse matrix of matrix V, respectively. This matrix V can be expressed by using the eigenvectors of the matrix C

$$V = \begin{pmatrix} v_{1,1} & v_{2,1} & \cdots & v_{N,1} \\ v_{1,2} & v_{2,2} & \cdots & v_{N,2} \\ \vdots & \vdots & \ddots & \vdots \\ v_{1,N} & v_{2,N} & \cdots & v_{N,N} \end{pmatrix} \tag{4.12}$$

where the k-th column of V is the k-th eigenvector $\boldsymbol{v}_k$ of a matrix C, such as

$$\boldsymbol{v}_k = \begin{pmatrix} v_{k,1} \\ v_{k,2} \\ \vdots \\ v_{k,N} \end{pmatrix} \tag{4.13}$$

The eigenvalues λ_k and the associated eigenvectors $\boldsymbol{v}_k$ of a matrix C satisfy the following equation:

$$C\boldsymbol{v}_k = \lambda_k \boldsymbol{v}_k \ (k = 1, \ldots, N) \tag{4.14}$$

Or equivalently in terms of matrix elements explicitly,

$$\begin{pmatrix} c_{1,1} & c_{1,2} & \cdots & c_{1,N} \\ c_{2,1} & c_{2,2} & \cdots & c_{2,N} \\ \vdots & \vdots & \ddots & \vdots \\ c_{N,1} & c_{N,2} & \cdots & c_{N,N} \end{pmatrix} \begin{pmatrix} v_{k,1} \\ v_{k,2} \\ \vdots \\ v_{k,N} \end{pmatrix} = \lambda_k \begin{pmatrix} v_{k,1} \\ v_{k,2} \\ \vdots \\ v_{k,N} \end{pmatrix} \tag{4.15}$$

By normalizing $\boldsymbol{v}_k$ such that

$$\boldsymbol{v}_k \cdot \boldsymbol{v}_k = \sum_{n=1}^{N} v_{k,n}^2 = 1 \tag{4.16}$$

the eigenvalue equation can also be written as

$$\boldsymbol{v}_l^t C \boldsymbol{v}_k = \lambda_k \delta_{k,l} \tag{4.17}$$

It can also be written explicitly with components as follows.

$$\begin{pmatrix} v_{l,1} v_{l,2} \cdots v_{l,N} \end{pmatrix} \begin{pmatrix} c_{1,1} & c_{1,2} & \cdots & c_{1,N} \\ c_{2,1} & c_{2,2} & \cdots & c_{2,N} \\ \vdots & \vdots & \ddots & \vdots \\ c_{N,1} & c_{N,2} & \cdots & c_{N,N} \end{pmatrix} \begin{pmatrix} v_{k,1} \\ v_{k,2} \\ \vdots \\ v_{k,N} \end{pmatrix} = \lambda_k \delta_{k,l} \tag{4.18}$$

Since the eigenvectors with different suffices k and l are orthogonal to each other,

$$\boldsymbol{v}_k \cdot \boldsymbol{v}_l = \sum_{n=1}^{N} v_{k,n} v_{l,n} = 0 \tag{4.19}$$

Using Kronecker's delta, Eqs. (4.16) and (4.19) can be written as

$$\boldsymbol{v}_k \cdot \boldsymbol{v}_l = \delta_{k,l} \tag{4.20}$$

The right-hand side of Eq. (4.20) is 0 for $k \neq l$ and is 1 for $k = l$. In matrix representation, the above process can be written as

$$V^t C V = \begin{pmatrix} v_{1,1} & v_{1,2} & \cdots & v_{1,N} \\ v_{2,1} & v_{2,2} & \cdots & v_{2,N} \\ \vdots & \vdots & \ddots & \vdots \\ v_{N,1} & v_{N,2} & \cdots & v_{N,N} \end{pmatrix} \begin{pmatrix} 1 & c_{1,2} & \cdots & c_{1,N} \\ c_{2,1} & 1 & \cdots & c_{2,N} \\ \vdots & \vdots & \ddots & \vdots \\ c_{N,1} & c_{N,2} & \cdots & 1 \end{pmatrix} \begin{pmatrix} v_{1,1} & v_{2,1} & \cdots & v_{N,1} \\ v_{1,2} & v_{2,2} & \cdots & v_{N,2} \\ \vdots & \vdots & \ddots & \vdots \\ v_{1,N} & v_{2,N} & \cdots & v_{N,N} \end{pmatrix}$$
$$= \begin{pmatrix} \lambda_1 & 0 & \cdots & 0 \\ 0 & \lambda_2 & \cdots & 0 \\ \vdots & \vdots & \ddots & \vdots \\ 0 & 0 & \cdots & \lambda_N \end{pmatrix} \tag{4.21}$$

is represented by the matrix V with the k-th eigenvector of matrix C as the k-th column when $\lambda_k \neq \lambda_l$ for $k \neq l$.

In practice, the eigenvalues of C can be obtained by Jacobi's diagonalization algorithm. The rest of the technical part of the diagonalization is the same as the Eq. (2.22)–(2.26) and Appendix B. The only difference of the RMT-test compared to the RMT-PCA is that the "i-th stock" for $i = 1, \ldots, N$ for the RMT-PCA is replaced by the "i-th block of a long data sequence" for $i = 1, \ldots, N$ for the RMT-test, in Fig. 4.1. The elements of the correlation matrix C in the case of the RMT-test are the inner products of the i-th block and the j-th block of the same data sequence, unlike the case of the RMT-PCA in Chaps. 2 and 3 where we took the inner products of price time series of the i-th stock and the j-th stock. The computation of the eigenvalues λ_i by diagonalizing the matrix C is in the same way by using the same program in Appendix C.

Once all the N eigenvalues are obtained, they are renamed in the descending order of magnitudes,

$$\lambda_N < \cdots < \lambda_3 < \lambda_2 < \lambda_1 = \lambda_{\max} \tag{4.22}$$

Those eigenvalues are used to compute a histogram of eigenvalue distribution and then compared with the theoretical curve given by the following equations:

$$P_{RMT}(\lambda) = \frac{Q}{2\pi\lambda}\sqrt{(\lambda_+ - \lambda)(\lambda - \lambda_-)} \tag{4.23}$$

$$\lambda_\pm = 1 + \frac{1}{Q} \pm 2\sqrt{\frac{1}{Q}} \tag{4.24}$$

where $Q = L/N$. They are obtained from the random matrix theory in the limit of N and L going to infinity by keeping the ratio $Q = L/N$ fixed. The randomness level is judged to be high if the histogram and the theoretical curve overlap and low if they do not overlap. This process is inevitably suffered by an error coming from the choice of the width to draw the histogram. For $N = 100$ or smaller, the number of partitions cannot be made larger than ten or so, for all the partitions filled by some data.

4.3 Qualitative Version

Naively, the data string is certified as "random," i.e., it passes the RMT-test, if the eigenvalue distribution of the correlation matrix C obtained by the real data coincides with the theoretically derived Marchenko-Pastur function (Marchenko and Pastur 1967; Sengupta and Mitra 1999) in Eqs. (4.23) and (4.24). A typical example is shown in Fig. 4.2. We can visually compare the probability distribution obtained from the histogram of the eigenvalue distribution to the RMT theoretical distribution function. The image can be depicted by comparing the left figure showing random and the right figure showing off-random in Fig. 4.2. The RMT-test in the left figure shows that the data used to produce the histogram "passes" the RMT-test, because the histogram and data used in the right figure are the case of less random data that fails the RMT-test. Incidentally, Q is a free parameter, and we choose $Q = 3$ for most cases in this article, including Fig. 4.2. By choosing $Q = 3$, the upper bound of the theoretical limit in Eq. (4.24) is $\lambda_+ = 2.488$, which makes the range of the figure reasonably large enough to visualize the distribution. Also, the case of $Q = 4$ and $\lambda_+ = 2.25$ is used later in Chap. 5. The values of Q larger than 4 make the distribution too dense to visualize. For example, the case of $Q = 6$ already makes the resolution

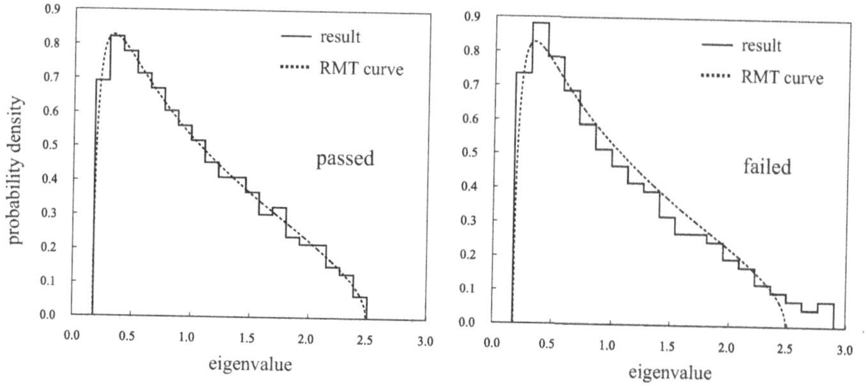

Fig. 4.2 Random data (left) passing and less random data (right) failing the RMT-test

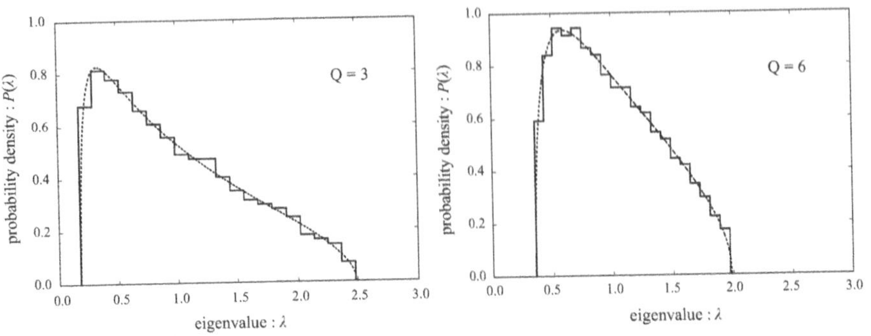

Fig. 4.3 Random numbers by LCG pass the qualitative RMT-test for $Q = 3$ and $Q = 6$

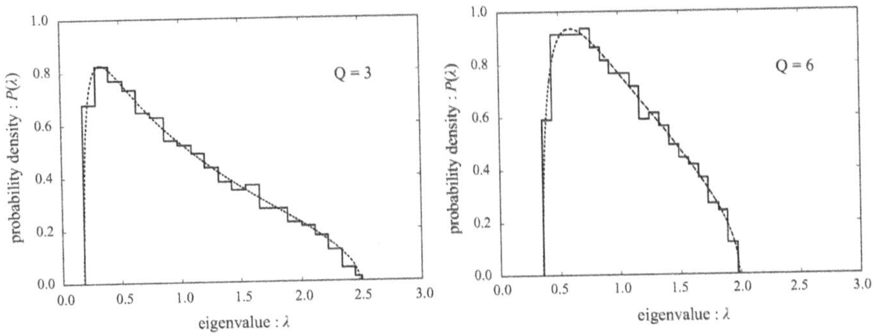

Fig. 4.4 Random numbers by MT pass the qualitative RMT-test for $Q = 3$ and $Q = 6$

of the figure around the neighbors of the threshold at $\lambda = \lambda_+$ too poor, as can be seen in Figs. 4.3 and 4.4 shown later.

First, we apply our method to general pseudo-random numbers and confirm that they pass the RMT-test using the method in Fig. 4.2. The candidates are linear congruential generator (LCG, abbreviated) (Knuth 1997) and Mersenne Twister (MT, abbreviated) (Matsumoto and Nishimura 1998); LCG is the most widely used and is already installed in many machines as the rand() function, popular with C programmers. Here, in order to control SEED, we dare to use the following function to fix the initial value of the output.

$$
\begin{aligned}
X_{n+1} &= (aX_n + b) \bmod M \\
a &= 1103515245, \\
b &= 12345, \\
M &= 2147483648
\end{aligned}
\tag{4.25}
$$

This simplicity is the strength of this algorithm; its weakness is its short period M. MT, on the other hand, has a very long period $2^{19937} - 1$, although the algorithm is not so simple. We used the source code downloaded from the inventor's home website (Matsumoto and Nishimura 1998). By using the data that we generated by means of those two generators LCG and MT, we computed the degree of randomness of LCG and MT and examined whether those two generators pass the randomness test.

Needless to say, LCG and MT passed the qualitative version of RMT-test, for $Q = 2, 3, \ldots, 10$, and $N = 100 - 500$, showing sufficient randomness levels for such a wide range of parameters as $N = 500$, $Q = 3$, and $Q = 6$, as shown in Fig. 4.3 for LCG and Fig. 4.4 for MT, respectively. So far, so good. Using this method, one can visually recognize randomness and distinguish between random and non-random sequences.

However, this method is not suitable for comparing the levels of randomness between highly random sequences, such as LCG and MT. Which is more random, and how random is it? If one seeks to compare subtle differences between two highly random sequences, one needs to quantify the method.

Also, the level of randomness in big data, such as tick-by-tick data for many stock prices, may be ridiculously low, since we would not expect rand() or MT to be as random as they are. If we want to use the randomness as a label for various big data, we need to quantify the method and assign a certain numerical tag to each piece of data.

4.4 Quantitative Version with Moments

The qualitative version of the RMT-test was validated for both LCG and MT. That is, with parameters $N = 500$ and $L = 1500$ (or 3000) and ratio $Q = L/N = 3$ (or 6), the numerical values of the eigenvalue distribution overlapped the theoretical curve of the RMT formula, confirming that the random numbers generated by the two algorithms are sufficiently random. The next task is to establish a quantitative version of the RMT-test to compare the subtle differences between the two random number generators. One way is to calculate the k-th moments defined as the average of the k-th power of all the eigenvalues as follows:

$$m_k = \frac{1}{N} \sum_{i=1}^{N} \lambda_i^k \tag{4.26}$$

Compare with the random limit calculated by

$$\mu_k = \int_{\lambda_-}^{\lambda_+} \lambda^k P_{RMT}(\lambda) d\lambda \tag{4.27}$$

In the quantitative version of the RMT-test, the degree of randomness is expressed by the deviation D_k between the moment m_k of the order k and its theoretical counterpart μ_k.

$$D_k = \frac{m_k}{\mu_k} - 1 \tag{4.28}$$

If both LCG and MT pass the qualitative RMT-test, then their degrees of randomness are indistinguishable by the qualitative test. In other words, the condition of "good randomness" can be defined by using the magnitude of the deviation rate D_k as follows.

$$|D_k| < \varepsilon = 0.02 \tag{4.29}$$

Since the case of $k = 1$ is trivial ($D_1 = 0$), we will show the k-th moment for $k = 2, \ldots, 6$. By choosing a range of order k and a suitable fixed small upper bound ε, e.g., $\varepsilon = 0.05$, this quantity acts as the inverse of the degree of randomness. Later, the magnitude of the deviation rate D_k is called the "error" and is sometimes referred to as the "randomness criterion."

The theoretical moments μ_k are obtained by substituting Eqs. (4.23) and (4.24) into Eq. (4.27). Explicit forms up to the 6-th order are as follows (Tanaka-Yamawaki et al. 2012b; Yang et al. 2013a).

$$\begin{aligned}
\mu_1 &= 1 \\
\mu_2 &= 1 + \frac{1}{Q} \\
\mu_3 &= 1 + \frac{3}{Q} + \frac{1}{Q^2} \\
\mu_4 &= 1 + \frac{6}{Q} + \frac{6}{Q^2} + \frac{1}{Q^3} \\
\mu_5 &= 1 + \frac{10}{Q} + \frac{20}{Q^2} + \frac{10}{Q^3} + \frac{1}{Q^4} \\
\mu_6 &= 1 + \frac{15}{Q} + \frac{50}{Q^2} + \frac{50}{Q^3} + \frac{15}{Q^4} + \frac{1}{Q^5}
\end{aligned} \tag{4.30}$$

The usage of moments saves us from the time-consuming diagonalization process of a huge matrix C. Using the trace theorem, we have

$$\sum_{i=1}^{N} (C^k)_{i,i} = \sum_{i=1}^{N} \lambda_i^k \tag{4.31}$$

Then, substituting for the moment in Eq. (4.26), we obtain the k-th moment

$$m_k = \frac{1}{N} \sum_{i=1}^{N} (C^k)_{i,i} \tag{4.32}$$

In other words, the moments in Eq. (4.30) are directly obtained by taking trace of the k-th power C^k of the correlation matrix C, without solving the eigenvalue problem to find all the eigenvalues. However, the program in Appendix C is designed to diagonalize the matrix to find all eigenvalues, since the largest eigenvalue λ_1 is required for computing $\Delta\lambda$, another randomness measure, to be introduced in the next section, convenient for processing a large amount of data with low randomness.

We now have all the necessary tools. We are in a position to determine the dimension N of correlation matrix, the length L of the data string, the degree of moment k needed to distinguish the subtle differences between two random sequences, and finally the optimal choice of parameters.

4.5 Highly Random Data

First, the quantitative version of the RMT-test is applied to the two pseudo-random generators, LCG and MT. These are used to determine the randomness criterion for discriminating good random numbers for an appropriate range of parameters k, N, and ε in the manner shown in Fig. 4.2.

In fact, the random number sequences generated by LCG and MT are not uniform. The level of randomness $|D_k|$ takes various values depending on the SEED, at which the algorithm starts the random number generation. Therefore, we prepared data by applying LCG and MT, starting from 100 different values of SEEDs, such as SEED$= 1, \ldots, 100$, then used the RMT test on those data individually to obtain the error $|D_k|$, and took average over 100 SEEDs, for $k = 2, \ldots, 6$. The results are shown in Fig. 4.5.

As N increases from 100 to 400, the error deviation from randomness, quantified by $|D_k|$, gradually deceases and becomes stable when $N = 500$ is reached. At $N = 500$, the mean values of $|D_k|$ are all smaller than 0.003. In addition, the upper limit ε can be evaluated by considering the maximum value of $|D_6|$, since $|D_k|$ increases with increasing k at least in the range of $k = 2, \ldots, 6$. Since the variation is twice the standard deviation, the upper limit ε of $|D_6|$ for data to be judged as random can be estimated as $\varepsilon = 0.02$. From these considerations, the value $N = 500$ is justified as being large enough to apply the RMT-test for all moments with $k = 6$, assuming $Q = 3$ and $\varepsilon = 0.02$.

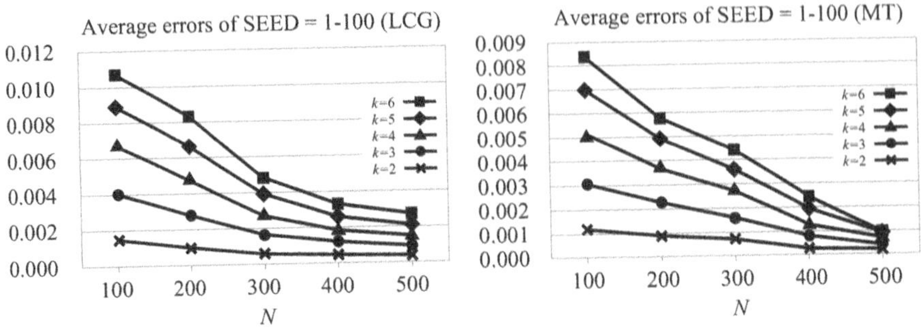

Fig. 4.5 Comparison of $|D_k|$ for various N for LCG and MT

Table 4.1 Mean (standard deviation) of $|D_k|$ for LCG and MT

k	LCG($Q = 3$)	MT($Q = 3$)	LCG($Q = 6$)	MT($Q = 6$)
2	0.0004(0.0010)	0.0004(0.0009)	0.0003(0.0006)	0.0001(0.0006)
3	0.0010(0.0026)	0.0009(0.0024)	0.0008(0.0016)	0.0004(0.0015)
4	0.0018(0.0047)	0.0014(0.0041)	0.0014(0.0028)	0.0006(0.0027)
5	0.0027(0.0072)	0.0019(0.0062)	0.0020(0.0043)	0.0009(0.0042)
6	0.0036(0.0100)	0.0022(0.0085)	0.0026(0.0060)	0.0012(0.0058)

For the later use, we shall define the randomness measure

$$|D_6| = \frac{m_6}{\mu_6} - 1 \tag{4.33}$$

as the deviation of the 6th moment m_6 defined in Eq. (4.29) or (4.32) from the RMT value μ_6 in Eq. (4.30). For example, $\mu_6 = 13.59671$ for $Q = 3$ and $\mu_6 = 8.71582$ for $Q = 4$, by Eq. (4.28).

The results of the quantitative version of the RMT-test for LCG and MT are summarized in Table 4.1. All values are the means of $|D_6|$ over 100 outputs corresponding to different SEEDs, and the standard deviation from the mean value for each result is given in the parentheses. The results can be summarized as follows.

(1) The sample data are random enough to pass the quantitative RMT-test, if $|D_6| < 0.02$ for $Q = 3$ and $N = 500$.
(2) The randomness of MT is generally higher than that of LCG.
(3) For $Q = 6$, the error is essentially smaller than for $Q = 3$.

The results so far show that the error between the experimental and theoretical moments does not converge to zero but gradually decreases to values of 0.2–0.4% or less in the region of N between 400 and 500. Is this due to the fact that the two pseudo-random numbers are not completely random or that the RMT formula is not completely valid for finite values of N and L?

To answer this question, we employed “physical random numbers” generated by a random number generator board installed in a supercomputer. Three types

Table 4.2 Comparison of $|D_6|$ for five types of random data ($Q = 3$)

	Pseudo random numbers by algorithm		Physically generated random numbers by boards		
k	LCG	MT	Hitachi1999	Toshiba2006	TED2010
2	0.0003 (0.0009)	0.0000 (0.0010)	0.0004 (0.0011)	0.0004 (0.0010)	0.0000 (0.0008)
3	0.0007 (0.0023)	0.0001 (0.0026)	0.0010 (0.0030)	0.0010 (0.0027)	0.0002 (0.0021)
4	0.0013 (0.0039)	0.0004 (0.0045)	0.0017 (0.0053)	0.0017 (0.0048)	0.0005 (0.0038)
5	0.0020 (0.0058)	0.0010 (0.0067)	0.0022 (0.0081)	0.0025 (0.0073)	0.0008 (0.0057)
6	0.0026 (0.0080)	0.0018 (0.0091)	0.0027 (0.011)	0.0033 (0.0101)	0.0010 (0.0079)

Table 4.3 Comparison of $|D_6|$ for five types of random data ($Q = 6$)

	Pseudo random numbers by algorithm		Physically generated random numbers by boards		
k	LCG	MT	Hitachi1999	Toshiba2006	TED2010
2	0.0001 (0.0005)	0.0003 (0.0006)	0.0004 (0.0010)	0.0001 (0.0006)	0.0002 (0.0005)
3	0.0003 (0.0014)	0.0007 (0.0016)	0.0010 (0.0021)	0.0004 (0.0016)	0.0005 (0.0013)
4	0.0006 (0.0026)	0.0012 (0.0030)	0.0015 (0.0035)	0.0008 (0.0029)	0.0009 (0.0025)
5	0.0008 (0.0041)	0.0017 (0.0046)	0.0021 (0.0051)	0.0012 (0.0045)	0.0012 (0.0038)
6	0.0010 (0.0059)	0.0021 (0.0065)	0.0028 (0.0069)	0.0016 (0.0063)	0.0014 (0.0053)

of random number generator boards were used: Hitachi1999, Toshiba2006, and TED2010, which were downloaded from the website of the Institute of Statistical Mathematics (Tamura 2010). The results of the analysis are summarized in Tables 4.2 ($Q = 3$) and 4.3 ($Q = 6$), showing the mean values of 50 independent experiments for the three physical random numbers and the standard deviation compared to LCG and MT in parentheses. Contrary to initial expectations, it turns out that the physical random numbers are not necessarily better random than LCG and MT. In particular, the average value of $|D_k|$, at $k = 2$ and 3, is comparable to that of LCG and MT. Thus, the advantage of physical random numbers is their unpredictability. In contrast, pseudo-random number generators such as LCG and MT operate deterministically and can reproduce the entire sequence if the SEED and algorithm are known. Among pseudo-random numbers, MT performs better than LCG; a look at the performance of the three physical boards shows technological progress, with the newest one (Tokyo ED) performing the best among the three, almost equal to MT. On the other hand, the older boards from Hitachi and Toshiba show almost the same performance as LCG.

4.6 Less Random Data: Measuring the Randomness by $\Delta\lambda = \lambda_1 - \lambda_+$

We need a good sample of data for testing the "low level of randomness." However, it is not a simple task to find good examples of low randomness data. By shuffling a regular sequence many times, we eventually reach the randomness. If we knew the inverse process of shuffling (un-shuffle), a less random data could be produced

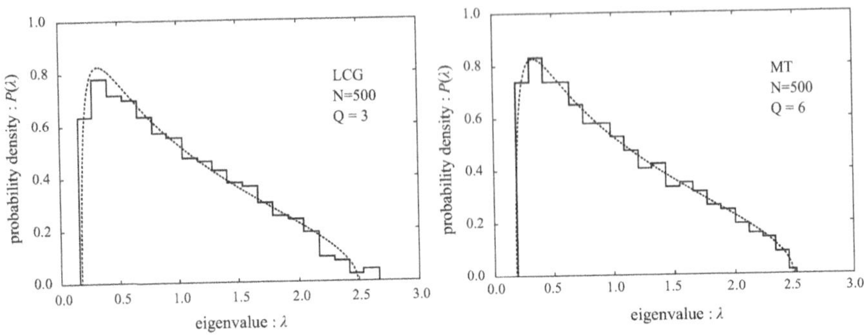

Fig. 4.6 The qualitative RMT-test for the initial 500 terms of LCG and MT

easily from the random sequence. Without such a process, it is rather hard to find a good example of not-so-random data around us.

It is known that the initial few hundred terms of LCG random numbers are less random than subsequent terms. It would be possible to use this to create sample data by collecting the initial portion of the LCGs immediately after the start of each generation in the various SEEDs. Based on this idea, we have made a test data by concatenating a set of the first 500 numbers generated by LCG starting from various SEEDs.

Figure 4.6 below shows the results of applying the qualitative version of the RMT-test to this test data. Clearly, we can see that the data collected from the initial portion of the LCG have larger errors than the previous results for the standard LCG and MT, whereas the data collected from the initial portion of the MT do not show such a trend. This result suggests that the randomness criterion in Eq. (4.29) can identify the initial portion of the LCG as less random than the MT or standard LCG. As numerical support for this observation, the results of applying the quantitative version of the RMT-test are shown in Table 4.4.

It can be seen that the initial part of the LCGs has a value of $|D_6|$ on the order of 0.06, which is about one order of magnitude larger than the standard LCGs shown in Table 4.4, for example. On the other hand, no such effect is observed in the initial part of MT.

Table 4.4 The deviation from the randomness $|D_6|$ for the initial part of LCG and MT

k	LCG ($Q = 3$)	MT ($Q = 3$)	LCG ($Q = 6$)	MT ($Q = 6$)
2	0.0045	0.0018	0.0057	0.0007
3	0.0103	0.0042	0.0141	0.0021
4	0.0197	0.0064	0.0251	0.0041
5	0.0352	0.0083	0.0392	0.0066
6	0.0583	0.0099	0.0571	0.0092

Another way to create less random data is to convert the highly random sequence $\{x_i\}$ to its log-return sequence $\{r_i\}$.

$$r_i = \log x_{i+1} - \log x_i = \log \frac{x_{i+1}}{x_i} \tag{4.34}$$

Incidentally, this log-return sequence is often used in finance to resolve differences between different stock prices. However, the time series obtained by this process has a special property. Even if x_i and x_{i+1} are independent, r_i and r_{i+1} are correlated, due to the overlapping of $\log(x_{i+1})$, as shown in Fig. 4.7. This type of overlapping can be avoided by using only the odd-numbered terms such as $(r_1, r_3, r_5, \ldots)$ or the even-numbered terms such as $(r_2, r_4, r_6, \ldots)$ at the cost of reducing the size of the original data to one-half. Figure 4.8 shows a result of following Fig. 4.7. The level of randomness is clearly reduced in Fig. 4.8 in comparison to Fig. 4.2 or Fig. 4.3.

The log-return data have much lower level of randomness compared to the initial part of the LCG. As shown in Table 4.5, the errors of log-return sequences have two

$$\begin{aligned} r_1 &= \log x_2 - \log x_1 \\ r_2 &= \log x_3 - \log x_2 \\ r_3 &= \log x_4 - \log x_3 \end{aligned}$$

Fig. 4.7 Randomness can be lowered by taking log-return

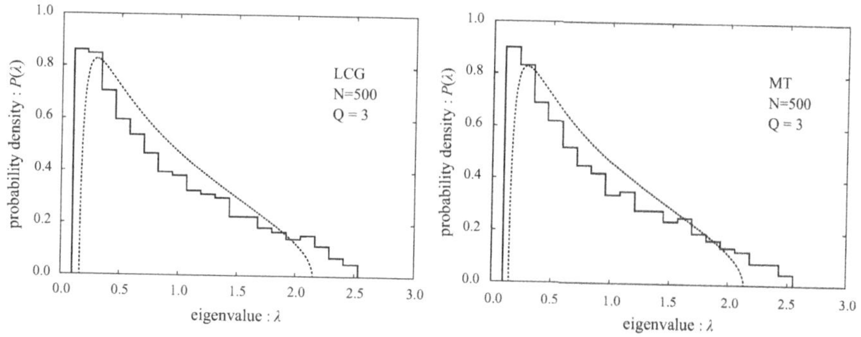

Fig. 4.8 The qualitative RMT-test for the log-return for LCG and MT

Table 4.5 The deviation $|D_6|$ for the log-return sequences

k	LCG($Q = 3$)	MT($Q = 3$)	LCG($Q = 6$)	MT($Q = 6$)
2	0.1047	0.1227	0.0696	0.0702
3	0.2578	0.3088	0.1866	0.1892
4	0.4445	0.5442	0.3391	0.3450
5	0.6596	0.8260	0.5240	0.5342
6	0.9092	1.1740	0.7426	0.7579

digits larger compared to the original, thus clearly failing the criterion $|D_6| < 0.02$ of the RMT-test.

The level of randomness can easily be judged by the qualitative version of the RMT-test, as shown in Fig. 4.2. The eigenvalue distribution of the correlation matrix made from data with high randomness overlaps with the theoretical distribution, as shown in the left figure. On the other hand, the eigenvalue distribution of the correlation matrix made from data with low randomness is shown on the right, indicating that there are eigenvalues that are larger than the maximum value λ_+ of the theoretical distribution. This line of thought leads us to use another index of randomness as the range of eigenvalues obtained from data, $\lambda_1 - \lambda_-$ compared to the range of eigenvalues in the limit of the RMT, $\lambda_+ - \lambda_-$

$$R = \frac{\lambda_1 - \lambda_-}{\lambda_+ - \lambda_-} \tag{4.35}$$

This value turns out to be approximately in the range of

$$R = 1.20 - 1.21 \tag{4.36}$$

as shown in Table 4.6, for the log-return time series of LCG as well as MT. This means that the range of eigenvalues $(\lambda_1 - \lambda_-)$ is approximately 20 % larger than the theoretical value of RMT, $R = 1$. This index R is another candidate for the index of randomness, in addition to the $|D_k|$.

More directly, the deviation from the randomness itself can be used as an index of deviation from the randomness, $\Delta\lambda$ defined by the following equation and the Fig. 4.9.

$$\Delta\lambda = \lambda_1 - \lambda_+ \tag{4.37}$$

In Fig. 4.9, $\Delta\lambda = \lambda_1 - \lambda_+$ is illustrated as the distance between the maximum eigenvalue λ_1 and the maximum λ_+ of the solid line. The dotted line shows the

Table 4.6 Range of eigenvalues for log-return data

Q	RMT range	LCG		MT	
	$\lambda_+ - \lambda_- = 4/\sqrt{Q}$	$\lambda_1 - \lambda_-$	$R = \frac{\lambda_1 - \lambda_-}{\lambda_+ - \lambda_-}$	$\lambda_1 - \lambda_-$	$R = \frac{\lambda_1 - \lambda_-}{\lambda_+ - \lambda_-}$
2	2.83	3.43	1.21	3.43	1.21
3	2.31	2.79	1.21	2.80	1.21
4	2.00	2.40	1.20	2.42	1.20
5	1.78	2.15	1.19	2.15	1.20
6	1.63	1.97	1.21	1.97	1.20
7	1.51	1.81	1.20	1.82	1.21
8	1.41	1.70	1.21	1.70	1.20
9	1.33	1.61	1.21	1.60	1.20
10	1.26	1.51	1.20	1.49	1.18

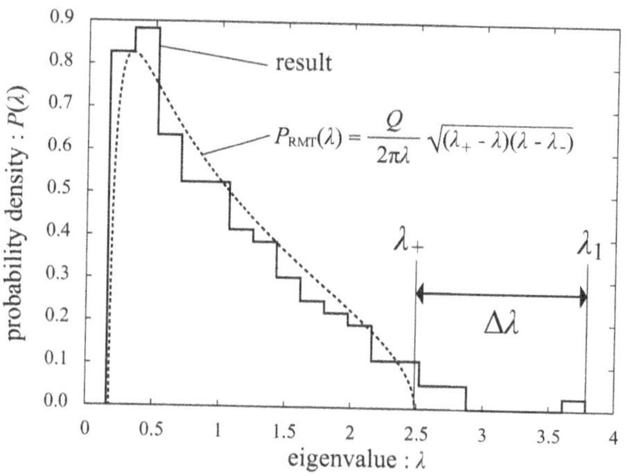

Fig. 4.9 The $\Delta\lambda = \lambda_1 - \lambda_+$ is illustrated

histogram of the eigenvalue distribution obtained from the data. The solid line shows the theoretical curve in the RMT limit.

Out of the three indices of randomness we have introduced, $|D_k|$, R, and $\Delta\lambda$, we prefer using $|D_k|$ for highly random data. On the other hand, R or $\Delta\lambda$ is more suitable for big data where the level of randomness is relatively low. The main advantage of D_k is twofold. One is the technical simplicity. We do not need the tedious process diagonalizing the matrix C. We can simply compute the trace (sum of the diagonal elements) of C^k as in Eq. (4.32) to compute the k-th moments. Secondly, we do not need to care the detailed description of the comparison between the k-th moment, m_k and its RMT version μ_k by using multiple orders k. On the other hand, the advantage of $\Delta\lambda$ or R for big data and other data of low randomness lies in the fact that the largest eigenvalue λ_1 is well-separated from the maximum edge λ_+ of the RMT spectrum of Eqs. (4.23) and (4.24). In this case, however, we cannot avoid solving the eigenvalue problem to obtain λ_1, which is a little more cumbersome to compute.

Summarizing the result obtained so far, the dimension of the correlation matrix N larger than or equal to 500 is desirable, and the length of each piece of the fraction of strings L larger than or equal to 1500 is necessary to determine the randomness level of data strings. Using those parameters, the data string passes the RMT-test at the same level of LCG, MT, or physical random numbers, if the condition expressed in Eq. (4.29) is satisfied. Namely, the randomness is as high as the well-known random number generators if $|D_6| < 0.02$ is satisfied, and the randomness is not high but the same level of the initial few hundred terms of LCG if $0.02 < |D_6| < 0.1$ is satisfied, and the randomness is as low as the log-return time series of a random sequence if $0.7 < |D_6| < 1.3$ is satisfied.

Now, we have so far used qualitative evaluation if the randomness is so low that it can be easily recognized by looking at the eigenvalue distribution diagram, but the method of quantifying the randomness by pseudo-random numbers and k-th moments is effective when the randomness is high.

However, for comparisons of random numbers with high randomness, such as pseudo-random numbers or physical random numbers obtained from the output of a random number board, which cannot be distinguished by qualitative evaluation, we have relied on quantitative evaluation by k-th moments. However, since the randomness of any random number varies, if k is too small, there will be no difference, so $k = 6$ was used. This implies, however, that as k is taken larger, the contribution of the largest eigenvalue stands out among the contributions of all eigenvalues to the moment, since the specific weight of the contribution of the larger eigenvalues increases. Therefore, when processing large amounts of data, it is not necessarily necessary to use moments, but rather it is convenient to use exponents for the new deviation of the largest eigenvalue λ_1 from the theoretical curve. The degree of randomness can be easily seen in the qualitative version of the RMT-test as shown in Fig. 4.2. Good random numbers behave as shown in the left figure, while non-random ones behave as shown in the right figure. In other words, when the data sequence is random, the maximum eigenvalue λ_1 is close to the maximum value of the theoretical curve λ_+, and when the data sequence is not random, the distance $\Delta\lambda = \lambda_1 - \lambda_+ = 1 - +$ increases. With this in mind, another (inverse) measure of randomness is the range of eigenvalues compared to the range of RMT, which gives the highest randomness.

4.7 Comparison to NIST

An independent criterion to compare with the RMT-test would help us to determine the objective border between the high randomness and low randomness. The NIST randomness test is expected to be a good candidate. On the home page of NIST, the National Institute of Standards and Technology, the NIST SP 800 is recommended as the US standard statistical randomness test. In addition, NIST randomness test was widely used to test the security level of ciphers. At the time of our experiment, the source code of NIST SP 800-22 was published on its website (NIST home page). According to this page, it is designed to test the randomness of data string by applying the 15 assays listed below.
List of 15 assays:

(1) Frequency (Monobit) Test
(2) Block Frequency Test
(3) Cumulative Sums
(4) Runs
(5) Longest Run
(6) Rank of matrix
(7) Fast Fourier Transform
(8) Non-overlapping Template
(9) Overlapping Template
(10) Universal (statistical assay by Maurer)

(11) Approximate Entropy
(12) Random Excursions
(13) Random Excursions Variant
(14) Serial
(15) Linear Complexity

We use them by setting the length L of one block for each assay as follows.

$L = 128$ in Assay (2)
$L = 9$ in Assay (8) and (9)
$L = 10$ in Assay (11)
$L = 16$ in Assay (14)
$L = 500$ in Assay (15)

The length of the sequence recommended is 1,000,000. In addition, at least 55 samples are to be prepared to obtain statistically significant results.

To determine the criteria to be highly random by comparing the NIST randomness test and the RMT-test, it is essential to prepare the data sequences of a fixed length (very large, e.g., 1,000,000) in various levels of randomness. As a matter of fact, however, it is rather hard to find data sequences having intermediate levels of randomness. For example, the randomness levels of the output sequences of pseudo-random generators installed in computers are too high, and the randomness levels of the real-world data given by financial, social, or biological statistics are too low in general. Therefore, it is more convenient to create a set of artificial data in various levels of randomness by shuffling a regular sequence. As the number of shuffles increases, the randomness of the sequence is expected to grow. For satisfying the requirement of the NIST, 55 sequences are prepared each of which has the length of 1,000,000. The initial sequence is constructed by placing 50 consecutive 0's and 50 consecutive 1's alternatively by 10,000 times to make the total length to be 1,000,000. This guarantees the equal frequencies of 0's and 1's in the data sequences. Then a shuffling is applied by exchanging randomly chosen two elements in the sequence. By repeating this shuffling T times, various data sequences are created. It is expected that the degree of randomness monotonically increases as T increases. For example, the sequences of one million shuffles and two million shuffles are constructed as follows, respectively.

(1) The initial sequence is shuffled by 1,000,000 times, and the result is written on the file named "the random number sequence of 1,000,000 shuffles."
(2) The sequence created in (1) is further shuffled by 1,000,000 times, and the result is written on the file named "the random number sequence of 2,000,000 shuffles."

By means of repeating this process, 55 samples are created for each of level of randomness. The number 55 is required to apply the NIST test.

Preparing those data sets, we have done the experiments as follows.

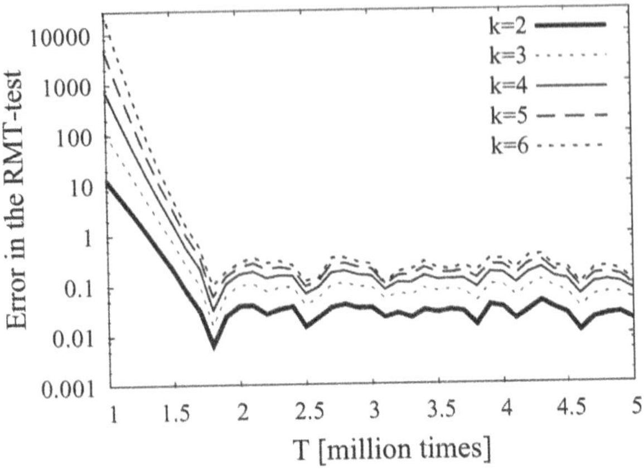

Fig. 4.10 The errors ($|Dk|$, k =2-6) decrease as T increases until reaching the saturation region

First, we confirm that the randomness indeed increases according to the number of shuffles. The result of applying the quantitative evaluation of the RMT-test on the random number sequences created by shuffling 1,000,000–5,000,000 times is summarized in Fig. 4.10, where the vertical axis shows the $|D_k|$ in Eq. (4.28) averaged over 55 samples, for the k-th moments of $k = 2, \ldots, 6$. The result shows the three facts.

(1) As the number of shuffles increases, the values of $|D_k|$ decrease monotonically before saturation region around $T = 1.7$ million. This fact confirms that the shuffling indeed helps increasing the randomness of given sequences.
(2) The error measured by $|D_k|$ reaches the minimum values, $|D_2|$=0.009 and $|D_6|$=0.1 around $T = 1.7$ million, and oscillates after that.
(3) The size of $|D_k|$ satisfies the inequality, $|D_k| < |D_{k'}|$ for $k < k'$.

To cross-validate the result, the NIST test is applied on the same set of data. Since the NIST is designed to apply on a sequence in ASCII format consisting of 0 and 1, the data sequences are converted to the $\{0, 1\}$ data by using the median as the boundary. For example, when the data consist of $\{00, 01, \ldots, 99\}$, $\{00, 01, \ldots, 49\}$ are assigned to $\{0\}$ and $\{50, 51, \ldots, 99\}$ are to $\{1\}$.

The result of the NIST randomness test using the sequences of 1,000,000–5,000,000 shuffles is summarized, together with the result of the RMT-test applied on the same set of data. It is shown that the data of $T = 1.7$ million times of shuffles pass at least 14 assays. In other words, both RMT-test and NIST reach saturation point at around $T = 1.7$ million times of shuffles.

The result can be summarized as follows.

(1) Both RMT-test and NIST detect the increase of randomness as T increases.
(2) Both RMT-test and NIST saturate after $T = 1.7$ million. After that point, both oscillate in a range.
(3) The criterion for NIST is to pass 14 or more assays and for the RMT is to satisfy $|D_6| \leq 0.6\%$.

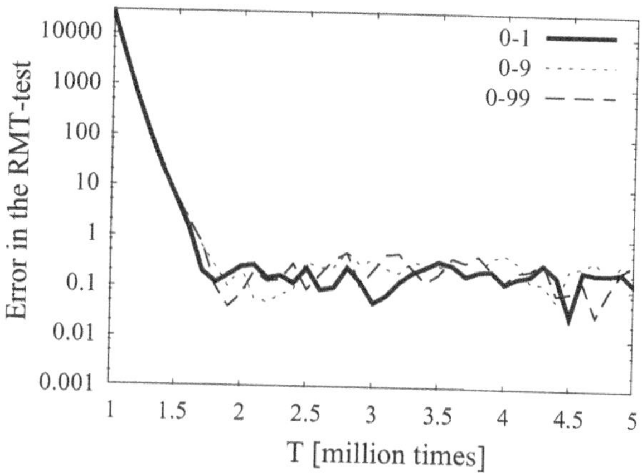

Fig. 4.11 Comparison of {0, 1}, {0, 1, ..., 9}, and {00, 01, ..., 99} in the ERROR as a function of T

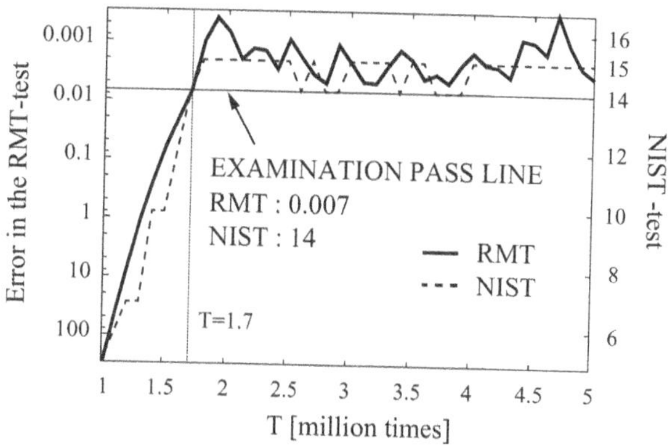

Fig. 4.12 Randomness measured by RMT-test and NIST increases by T

Now the next task is to check if the result depends on the number of symbols or not. Since the result obtained so far is on the case of the number of symbols being two {0, 1}, data using various kinds of symbols are used to check whether he result depends on the symbol or not. The initial sequence is prepared by placing ten decimal numbers {0, 1, ..., 9} in ascending order and combine ten thousand pieces of the same sequences to make the total length to be 1,000,000. Likewise, the 100 symbols {00, 01, ..., 99} are also used to make the total length 1,000,000. By means of shuffling these initial sequences by T times, data sequences having various levels of randomness are created. A comparison of the results of the RMT-test for three kinds of symbols is shown in Fig. 4.11.

In addition, the corresponding results of the NIST are shown in Fig. 4.12, in which the vertical axis in the right represents the number of passed assays out of total 15 in the NIST test, together with the result of RMT-test in the left axis showing the deviation from the randomness in terms of the 6th order moments $|D_6|$ by error. Details of which assay did not pass the NIST-test where in the case of NIST result is 14/15 are shown in Table 4.7.

Table 4.7 Comparison of the results of the RMT-test and the NIST randomness test

Shuffles $\times 10^4$	RMT $\|D_6\|$	NIST pass rate	Assay failed	Shuffles $\times 10^4$	RMT $\|D_6\|$	NIST pass rate	Assay failed
100	29359	5/15	2) 4) 5) 7) 8) 9) 10) 11) 12) 14)	310	0.23	15/15	
110	3801	7/15	2) 4) 5) 7) 8) 9) 11) 14)	320	0.22	15/15	
120	574	7/15	2) 4) 5) 7) 8) 9) 11) 14)	330	0.22	15/15	
130	105	7/15	4) 5) 7) 8) 9) 11) 13) 14)	340	0.20	14/15	8)
140	23	10/15	4) 7) 8) 9) 11)	350	0.19	15/15	
150	6	10/15	4) 5) 7) 8) 9)	360	0.18	14/15	8)
160	2	12/15	4) 7) 8)	370	0.14	14/15	8)
170	0.69	14/15	8)	380	0.14	14/15	8)
180	0.49	14/15	8)	390	0.14	14/15	8)
190	0.48	15/15		400	0.13	15/15	
200	0.46	15/15		410	0.13	15/15	
210	0.46	14/15	8)	420	0.11	15/15	
220	0.40	15/15		430	0.10	15/15	
230	0.35	15/15		440	0.10	15/15	
240	0.35	15/15		450	0.09	15/15	
250	0.35	15/15		460	0.09	15/15	
260	0.33	14/15	8)	470	0.08	15/15	
270	0.29	15/15		480	0.07	15/15	
280	0.24	14/15	8)	490	0.04	15/15	
290	0.24	14/15	8)	500	0.03	15/15	
300	0.24	15/15					

Table 4.8 Comparison of LCG and three physical random generators

Random generator	RMT-test $\|D_6\|$%	NIST test Pass rate
LCG	0.2831	14/15
Hitachi	0.1597	15/15
Toshiba	0.0026	15/15
Tokyo	0.1194	15/15

Finally, the cases of very high randomness given by the pseudo-random generators and physical random numbers are examined for the sake of comparison. As shown in Table 4.8, the values of |Error| for all the cases are less than 0.3% in the RMT-test, and the rate of passed assays is 14 or larger out of 15 in the NIST. We can conclude that after $T = 1.7$ million times of shuffles, the binary sequence reaches the saturation point, over which no more randomness is obtained. There the $|D_6|$ of the RMT-test ($k = 6$) becomes less than 0.3%, and 14 or more assays are satisfied in the NIST.

According to the survey, the sequences shuffled between 1.7 million and 1.8 million times failed only in the assay of "Runs," while the sequences more than 1.8 million times failed only in the assay of "Non-Overlapping Template." This fact seems to imply that the randomness calculated by the RMT- test is not related to the assay of "Non-Overlapping Template" in the NIST test.

To this end, we recall that the RMT test tries to detect correlations between the elements of time series, so is it completely sufficient to detect correlations to examine the degree of randomness. Quite simply, it is difficult to define which is the broader concept of randomness or uncorrelatedness. Is there an event that is uncorrelated but not random? Or is there an event that is random but correlated? No, but we hesitate to answer immediately if asked they are the same. Since NIST is designed to test randomness from multiple angles with 15 tests, the fact that it was always the "non-overlapping templates" that remained when only 14 out of 15 passed in our experiments may suggest a difference between the two methods.

Chapter 5
Applications of the RMT-Test

Next, we will discuss three applications of RMT-test that we have tried. The first application measures the randomness of two Hash functions. Naturally, the newest algorithm produces more random Hash values than the older one. But let's check: the second application is to select stocks to invest in; the third application is to use the randomness measured by the RMT-test to predict possible market disruptions in advance.

5.1 Hash Functions, MD-5 and SHA-1

In this section, we test the validity of the RMT-test using real data. As a first example, we compare the output data of two cryptographic hash functions, MD5 (Black et al. 2008) and SHA-1 (Locktyukhin and Farrel 2010). A cryptographic hash function is an algorithm that converts given data (a message) into a fixed-size sequence of bits called a cryptographic hash value. The hash value is also called a message digest or digest. New algorithms have been devised to solve the shortcomings of older algorithms, and SHA-1 and MD5 are now somewhat outdated algorithms, but for the purpose of checking the performance of RMT-test, it should be sufficient to compare two different types with different levels of randomness of output values. The basic characteristics of the two algorithms are summarized in Table 5.1.

Since SHA-1 came after MD5 in chronological order, the hash value of SHA-1 is expected to be more random than the older MD5. This is because hash values should not give any hint of the original message, and a good hash function is expected to produce a highly random output for the purpose of hiding the original message. Therefore, we attempted to show that SHA-1 is superior to MD5 by comparing the randomness of the hash values of these two algorithms through RMT-test.

M. Tanaka-Yamawaki, Y. Ikura, *Principal Component Analysis and Randomness Test for Big Data Analysis*, Evolutionary Economics and Social Complexity Science 25, https://doi.org/10.1007/978-981-19-3967-9_5

Table 5.1 Comparison of regulations of two hash functions, MD5 and SHA-1

	MD5	SHA-1
Date published	1992	1995
Structure	Merkle-Damgård construction	Merkle-Damgård construction
Digest style	Hexadecimal	Hexadecimal
Block of data	512	512
Length of digest	128	160
Number of steps between collisions	64	80

Table 5.2 Comparison of MD5 and SHA-1 by means of $|D_6|$

k	MD5($N = 200$, $Q = 3$)	SHA-1($N = 200$, $Q = 3$)
2	0.0017 (0.003)	0.0003(0.002)
4	0.0074 (0.014)	0.0005(0.007)
6	0.0137 (0.029)	0.0011(0.015)

For the two hash algorithms, MD5 and SHA-1, the absolute value of the 6-th moment, $|D_6|$, is shown in Table 5.2, as the mean value from ten independent experiments and the standard deviation in parentheses. Note that the smaller the value of $|D_6|$, the higher the randomness. First, we see that the average values over ten independent results of $|D_6|$ for both SHA-1 and MD5 are smaller than 0.05; thus both pass the RMT-test. Then we notice that the values of $|D_6|$ for SHA-1 are consistently smaller than those for MD5 by more than factor 5, although the standard deviations are rather large. Moreover, the qualitative version of the RMT-test in Fig. 5.1 shows a better fit of the histogram to the theoretical curve in the case of SHA-1, compared to the case of MD5.

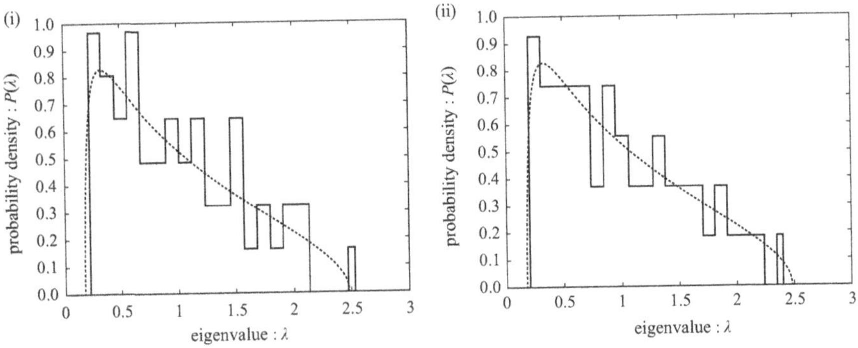

Fig. 5.1 Comparison of MD5 (left) and SHA-1(right) in the qualitative RMT-test

5.2 Discovering Safe Investment Issues Based on Randomness

How to choose safe stocks is a central problem in the stock investment. It would be exciting if one can use high-frequency price data to predict the future move of each stock based on a simple indicator. It would be worth trying our new candidate, the randomness by the RMT-test. This line of thought leads us to survey the effect of the RMT-test on the stock prices. We choose the second application of the RMT-test be the stock prices. As expected, the price time series would have much lower level of randomness compared to the output of hash functions or various random generators. Furthermore, it is customary to convert the price time series $S_i = (S_{i,1}, S_{i,2}, \ldots, S_{i,L+1})$ to the log-return time series $r_{i,1}, r_{i,2}, \ldots, r_{i,L}$ by means of Eq. (3.5),

$$r_{i,k} = \log S_{i,k+1} - \log S_{i,k}$$

where $S_{i,k}$ is the k-th price data of the i-th stock as given in various databases. In terms of discrete variables, the correlation $C_{i,j}$ between the return of two stocks, i and j, can be written as the inner product of the two log-return time series,

$$C_{i,j} = \sum_{k=1}^{L} g_{i,k} g_{j,k}$$

where $g_{i,k}$ is the normalized time series to have the zero averages and the unit variances as follows.

$$g_{i,k} = \frac{r_{i,k} - \langle r_i \rangle}{\sigma_i} \quad (i = 1, \ldots, N,\ k = 1, \ldots, L)$$

where

$$\langle r_i \rangle = \frac{1}{L} \sum_{k=1}^{L} r_{i,k}$$

and

$$\sigma^2 = \langle r_i^2 \rangle - \langle r_i \rangle^2$$

For the analysis, TOPIX 500 stock prices for every minute from 2007 to 2009 were used. Since trading does not necessarily take place every minute, a

Table 5.3 Data used in the analysis, year, data length (T), and number of stocks (M)

Year	Data length (T)	Number of stocks (M)
2007	66338	211
2008	66338	240
2009	65945	229

preprocessing step is necessary to prepare the N-by-L rectangular matrix of data G for the analysis. This process consists of the following three steps:

(1) Calculate the degree of randomness using the minute-by-minute price data for each stock each year. If a price is missing because there are no transactions every minute, the price of the previous minute is copied and filled in. However, stocks for which the price at which the missing price is filled by such an operation is more than 20% of the data length T for the year are excluded from the analysis. This screening process determines the parameter M in Table 5.3.
(2) The stock price time series $p_1, p_2, \ldots, p_T$ are transformed into log-return time series $r_1, r_2, \ldots, r_{T-1}$ by Eq. (4.34), as is usually done in financial analysis, in order to eliminate the unit and size dependence that differs for each stock price.
(3) From the data sequence of length T in each year for each stock code, an N-by-L matrix G is constructed by cutting the sequence into N pieces of equal length $L = NQ$. If T is not divisible by L, discard the remainder, construct the correlation matrix $C = GG^t$, and find the eigenvalues and eigenvector. Although Q is a freely chosen parameter, Figure $Q = 4$ is chosen here, which provides a good separation between high (H) and low (L) randomness stocks, as shown in Fig. 5.2. As shown in Table 5.3, $T = 66338$ for 2007 and 2008, and $T = 65945$ for 2009. In both cases, $N = 128$ and $L = 512$.

Our numerical experiment revealed in Chap. 4, $N \geq 100$, is preferred to measure the randomness of machine-generated random numbers, such as LCG or MT. In the real-data application, the data size cannot be chosen arbitrary. In our case of stock prices per 1 min, the data length for 2007 is 66338 according to Table 5.3. Although the parameter Q can be chosen arbitrary, preliminary experiments hinted us the choice of $Q = 4$ which most clearly shows the superiority of H compared to L throughout the next year as shown in Fig. 5.2. Choosing $Q = 4$, $N = 128$ and $L = 512$ are available to satisfy $66338 > NL = QN^2$. By setting this parameter, we can compute the randomness, by the deviation $|D_6|$ for each of 211 stocks in Table 5.3.

(1) Compute the randomness of stocks by the RMT-test.
(2) Sort the stocks according to the randomness to choose the first one and the last one.
(3) Investigate the profit (log-return) of each stock in the next year.
(4) Analyze the relationship between the profit and the randomness of stocks.

The result is as follows. The ranking of randomness in 2007 is shown in Table 5.4, in which the stock of the highest randomness in 2007 is 9504 in the

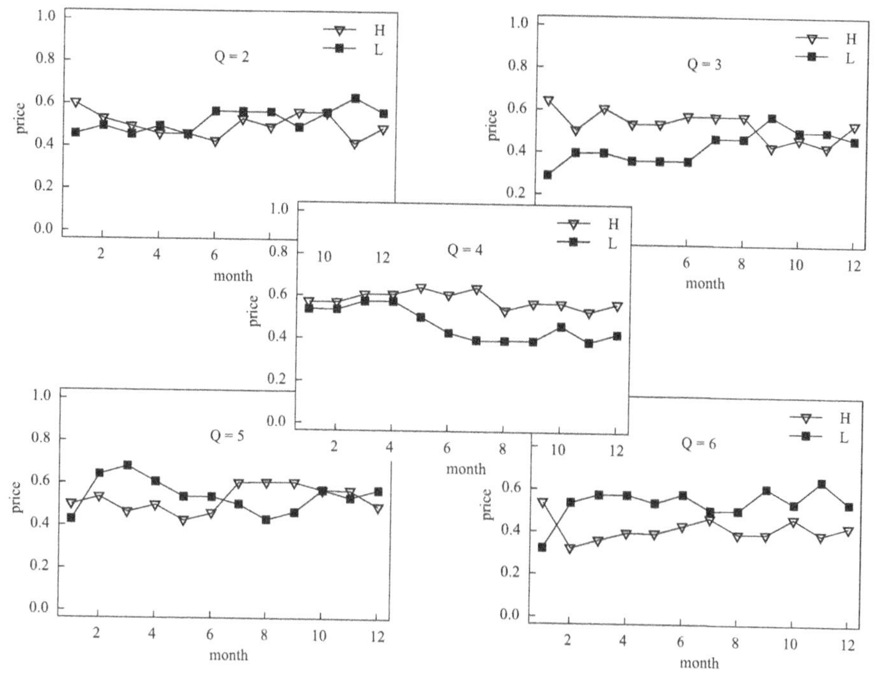

Fig. 5.2 The choice of $Q = 4$ is based on the better presentation of L<H next year

Table 5.4 The ranking of randomness by using the tick data of 2007

Rank	Sector	Code	$\|D_6\|$
1	Electric/Gas	9504	26.4
2	Machinery	6460	37.6
3	Electric/Gas	9506	38.2
4	Electric/Gas	9508	43.3
5	Information & Communication	4676	44.9
207	Electric appliances	6506	740.9
208	Nonferrous metals	5802	797.3
209	Chemicals	4043	799.8
210	Iron and steel	5541	1001.5
211	Transportation equipment	7201	1209.6

sector of Electric/Gas and of the lowest randomness in 2007 is 7201 in the sector of Transportation and Equipment. Figure 5.3 compares the log-return of those two extremes together with Nikkei Average. As a result, the most random stock performs better than the other stocks of the lower randomness and is affected very little by the financial crisis. On the contrary, the 7201 which has the lowest randomness in 2007 fell enormously by the effect of Lehman shock that occurred in September 2008. Based on this observation, the authors noticed that the stock which has the highest randomness is stable and safe under a bear market.

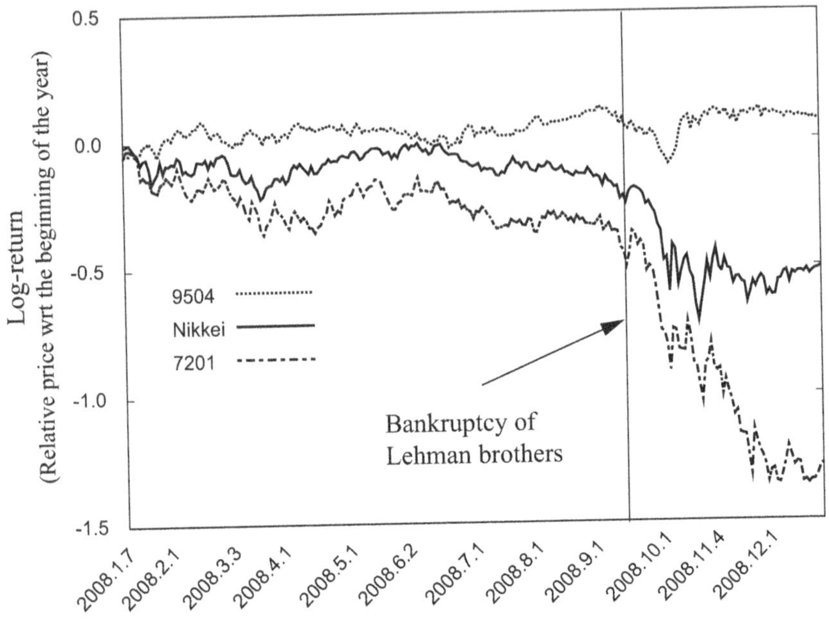

Fig. 5.3 The most/least random stock 9504/7201 in 2007 performs the best/worst in 2008

Likewise, the same rule holds for the relationship between the randomness in 2009 and the performance in 2010. From this observation, we can speculate an empirical rule "high randomness means low risk." However, the situation is opposite for the combination of the randomness in 2008 and the performance in 2008.

The ranking of randomness in 2008 is shown in Table 5.5, where the stock of the highest randomness is 8308 that belongs to the sector of Banks, which is not a stable stock. In other words, the empirical rule "high randomness means low risk" does not seem to work in 2008. The reason for this may be viewed as the effect of the Lehman shock occurred in September 2008, which made the rest of this year the period of an abnormal fluctuation of stock price. By amputating the abnormal part and using the data just before the end of August in 2008, it is expected to have the normal condition.

The result shown in Table 5.6 indeed shows the normal features, where the stable stock 9506 in Electric/Gas sector is extracted as the stock of the highest randomness, while a relatively unstable stock 7201 in Transportation and Equipment sector shows the lowest randomness. Because 7201 fell enormously by the effect of Lehman shock, it is very likely to rise in 2009. As a result, as shown in Fig. 5.4, 7201 began to rise in March 2009. This is a good example of a high profit accompanied by a high risk. On the other hand, almost of the top five highest randomness stocks perform better than the lowest randomness stock 7201 but 6728. It means that there will be some noise by using the RMT-test to judge the relationship between the randomness

Table 5.5 The ranking of randomness by using the tick data of 2008

Rank	Sector	Code	$\lvert D_6 \rvert$
1	Banks	8308	28.4
2	Machinery	7004	30.9
3	Transportation equipment	7211	31.8
4	Electric/Gas	9502	32.3
5	Electric/Gas	9508	36.3
236	Electric appliances	4902	1604.0
237	Electric appliances	6954	1611.2
238	Electric appliances	7752	1646.9
239	Securities	8604	2059.9
240	Shipping	9104	2097.5

Table 5.6 The ranking of randomness by using the tick data of Jan–Aug, 2008

Rank	Sector	Code	$\lvert D_6 \rvert$
1	Electric/Gas	9506	11.6
2	Electric appliances	6728	13.1
3	Foods	2267	19.1
4	Electric/Gas	9502	19.1
5	Retail trade	2685	22.2
230	Electric appliances	6762	183.0
231	Electric appliances	6503	183.7
232	Banks	8306	190.0
233	Securities	8604	222.8
234	Transportation equipment	7201	257.9

and log-return of stock; however, the empirical rule, "high randomness means low risk," is justified.

The 2007 results and the 2008 pre-Lehman Brothers data show that stocks with the highest randomness perform better than those with the lowest randomness, while those with the highest randomness are less affected when there is a financial crisis. The 2009 results are shown in Table 5.7 and Fig. 5.5, which shows that the stock with the code number 9509 in the Electricity and Gas sector was identified as the most random and is safer than the stock with the code number 8058; in the following year, most of the top five stocks with the highest randomness performed better than 8058 stocks in the wholesale sector with the lowest randomness. Thus, the rule of thumb that "more randomness equals less risk" held true for the stock market in 2009.

In other words, the degree of randomness in each year, measured by selecting approximately 220 stocks that satisfy the simultaneity condition and other conditions out of the 1-minute data of the 483 shares listed in the TOPIX500 for 2007–2009, is related to the stock price trend in the following year, indicating that the higher the degree of randomness, the safer the stock is in the following year.

However, since this is a period of falling stock prices as shown in Fig. 5.6, it can be interpreted that in fact, stocks with low liquidity and which did not move much

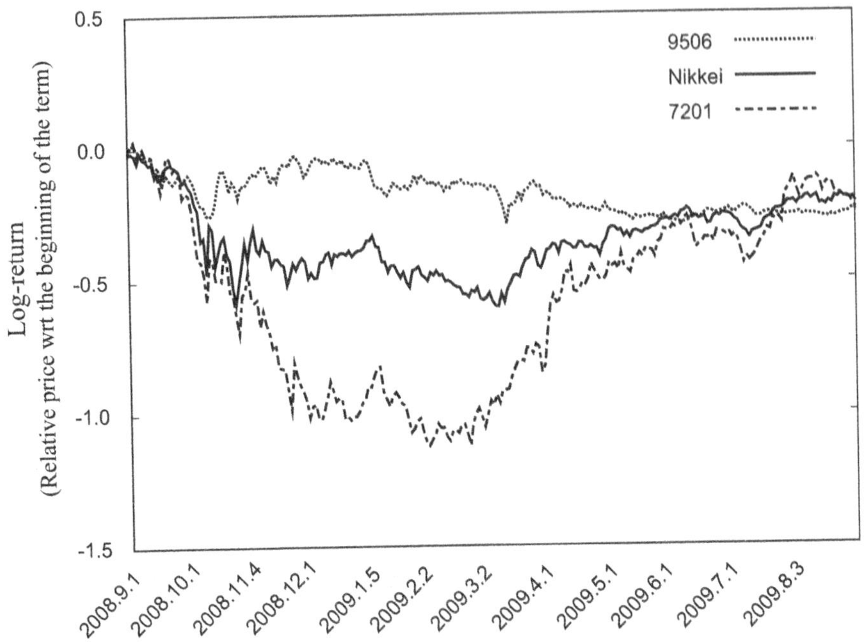

Fig. 5.4 Stock 9506 with the highest randomness outperformed stock 7201

Table 5.7 The randomness ranking by using the tick data of 2009

Rank	Sector	Code	$\lvert D_6 \rvert$
1	Electric/Gas	9508	25.5
2	Electric/Gas	9509	28.1
3	Electric/Gas	9506	31.6
4	Electric/Gas	9502	33.2
5	Retail trade	2651	36.4
225	Nonferrous metals	5713	1039.5
226	Electric appliances	4902	1072.4
227	Machinery	6301	1090.9
228	Iron and steel	5541	1128.2
229	Wholesale trade	8058	1249.5

are the safest stocks, while those with high trading volume and large movements are the least safe. If this is the case, it is not surprising that the opposite should happen in a so-called bull market. In order to confirm whether this is actually the case, we need data from periods when stock prices were rising. We searched for data to measure the randomness level of price fluctuation in a rising market.

Since data were limited at the time this experiment was conducted, we attempted to verify this using the NYSE tick data from 1993 to 2002, which were used in the first study in Chap. 3 and was named as Data A in Table 2.2, for the first 3 years of an upward trend in this period, i.e., 1993–1996. Figure 5.7 shows that the stock

Fig. 5.5 The most random stock 9508 outperforms the least random stock 8058

prices were in an uptrend during this period. In Chap. 3 when the same data was used for RMT-PCA, the correlation of many stock prices at the same time was required, and representative values for each hour would have to be extracted, yielding only inaccurate results. However, since the randomness was measured for each stock price, there is no need to consider the simultaneity among stock prices. This fact is quite helpful in calculating the magnitude of randomness for each year.

Tick data is time series data in which the stock price is recorded for each trade of a stock. In other words, tick data does not record prices during the time when there is no trading, so the length of the data differs for each stock. However, even for stocks with a small amount of data, we want to ensure that the amount of data is sufficient to evaluate the randomness of each stock each year in the RMT test and to standardize the parameters used in the validation for each year. Therefore, the annual data length for each stock is standardized by processing the tick data into 1-minute data. The process of converting tick prices to 1-minute data is illustrated in Fig. 5.7. Specifically, the price at 00 seconds or the price immediately before is used for each minute to record "XX:XX:00." For example, as shown in Fig. 5.8, there is no data for 15:59:00, but the price at the immediately preceding 15:58:32, \$65, corresponds to the current price. Since the market closes at 4:00 p.m., the price at 9:30 a.m., when the market opens the next morning, corresponds to the previous \$65.25, and the price at 9:31 a.m. corresponds to the \$64.75 immediately before that. In the case of a stock split, the price is also corrected. For example, in the case of a "3 for 1" stock split, the pre-split price is treated as one-third of the post-split

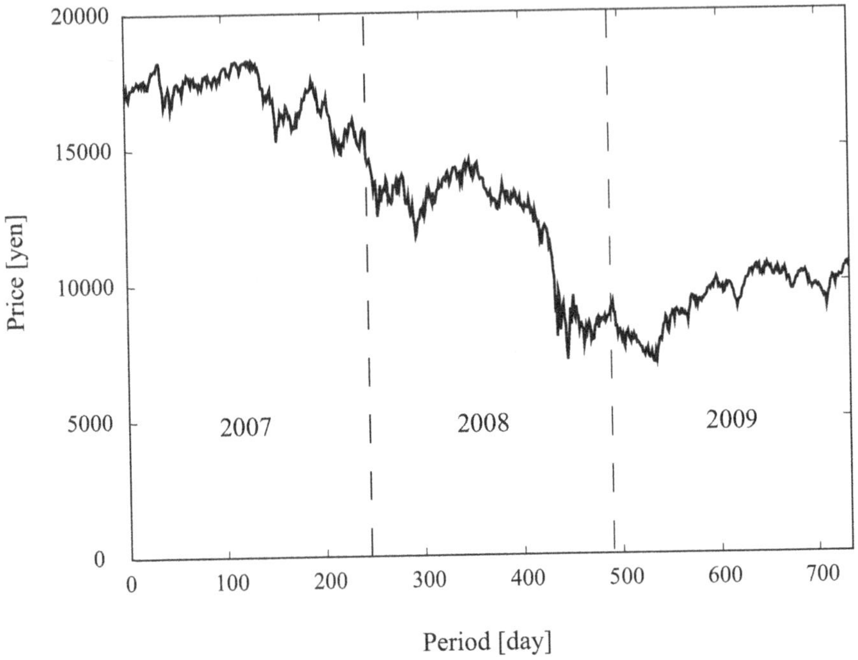

Fig. 5.6 The stock index Nikkei average in Japan was in the falling trend during the years 2007–2009

price and adjusted to match the post-split price. The daily trading hours on the New York Stock Exchange is from 9:30 to 16:00, for a total of 390 min, with some half-day business days. The number of working days varies year to year. In 1993, for example, the total length of the prices is 98743, and then by converting the prices to the log-return time series $r_1, r_2, \ldots, r_L$ by means of Eq. (3.5) which reduces the data length to 98742, from which we use the maximum length $T = NL = L^2/Q$ by discarding the beginning part in that year, we have $L = 624$ and $N = 157$, as in Table 5.8.

Using those data, assuming that they are properly normalized by Eqs. (4.2)–(4.4) for each stock code for each year, an N-by-L rectangular matrix G in Eq. (4.5) is constructed for each stock code in each year, to further construct a correlation matrix C. The degree of randomness is quantified by means of the deviation from randomness $|D_6|$ defined in Eq. (4.28), with smaller $|D_6|$ indicating greater randomness. The process of the job-flow of computing the randomness in the RMT test is illustrated in Fig. 5.9.

The results for 1993, 1994, 1995, and 1996 are shown in Tables 5.9, 5.10, 5.11, and 5.12, respectively.

Note that the values of the absolute errors of the sixth-order moments for the top and bottom stocks show that the error for the top stocks is 0.030 and the error for the bottom stocks is 34.36, indicating a large difference between the two. The

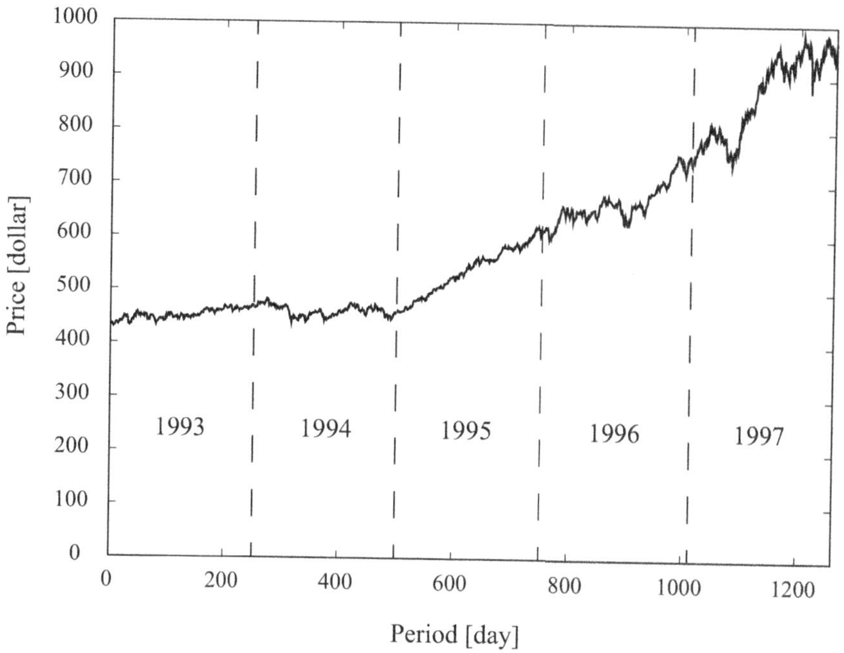

Fig. 5.7 The stock index S&P500 in USA was in the rising trend during the years 1993–1997

Table 5.8 Sizes of 1-minute price data in each year in 1993–1996

Year	1993	1994	1995	1996
Number of stock codes	222	244	250	263
Data points (price)	98743	98262	98112	98564
N	157	156	156	156
L	628	624	624	624

day	time	price [$]
...	...	...
Jan 3	15:58:32	65.00
Jan 3	15:59:43	64.75
Jan 3	16:00:00	64.50
Jan 3	16:00:31	65.25
Jan 4	9:30:44	65.00
Jan 4	9:30:57	64.75
Jan 4	9:31:09	64.75
...	...	...

⇨

day	time	price [$]
...	...	...
Jan 3	15:59:00	65.00
Jan 3	16:00:00	64.50
Jan 4	9:30:00	65.25
Jan 4	9:31:00	64.75
...	...	...

Fig. 5.8 Process of converting tick data to 1-minute data

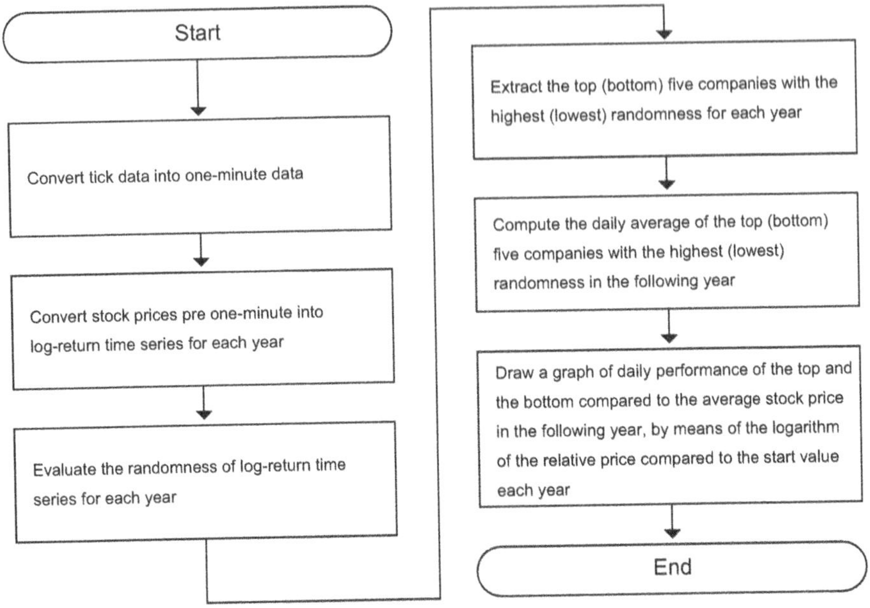

Fig. 5.9 Procedure of comparing randomness and future performance by RMT-test

Table 5.9 The deviation from randomness $|D_6|$ for 222 stocks in 1993 is shown in the descending order of randomness, ascending order of $|D_6|$

Rank	Sector	Ticker symbols	$\|D_6\|$
1	Finance	AMP	0.030
2	Material	BLL	0.072
3	Material	IFF	0.073
4	Information Tech	CSC	0.076
5	Consumer goods, service	WHR	0.095
218	Living essential	PEP	29.35
219	Telecommunication	T	29.73
220	Living essential	MO	31.66
221	Living essential	WMT	32.05
222	Health care	BMY	34.36

bottom five companies are dominated by popular companies such as PepsiCo (PEP), AT&T(T), Altria Group (MO), Wal-Mart Stores (WMT), and Bristol-Myers Squibb (BMY).

The annual stock performance is graphed by the log-return $E(n, k)$ which is calculated by taking the log-difference between the opening price $S(n, 0)$ of the term compared to the daily stock closing price $S(n, k)$ of the k-th day.

$$E(n, k) = \ln \frac{S(n, k)}{S(n, 0)} \tag{5.1}$$

Table 5.10 The deviation from randomness $|D_6|$ for 244 stocks in 1994 is shown in the descending order of randomness, i.e., ascending order of $|D_6|$

Rank	Sector	Ticker symbols	$\lvert D_6 \rvert$
1	Capital goods, service	ETN	0.091
2	Capital goods, service	DHR	0.18
3	Consumer goods, service	MAT	0.21
4	Living essential	KMB	0.23
5	Health care	BDX	0.24
240	Living essential	PEP	22.00
241	Capital goods, service	GE	22.01
242	Health care	MRK	28.14
243	Telecommunication	T	33.67
244	Capital goods, service	RHI	45.76

Table 5.11 The deviation from randomness $|D_6|$ for 250 stocks in 1995 is shown in the descending order of randomness, ascending order of $|D_6|$

Rank	Sector	Ticker symbols	$\lvert D_6 \rvert$
1	Finance	AFL	0.012
2	Public service	TE	0.013
3	Capital goods, service	HON	0.037
4	Capital goods, service	PH	0.047
5	Energy	HP	0.10
246	Living essential	WMT	21.00
247	Health care	MRK	21.13
248	Consumer goods, service	F	23.93
249	Telecommunication	T	24.35
250	Capital goods, service	GE	27.94

Table 5.12 The deviation from randomness $|D_6|$ for 263 stocks in 1996 is shown in the descending order of randomness, ascending order of $|D_6|$

Rank	Sector	Ticker symbols	$\lvert D_6 \rvert$
1	Consumer goods, service	MAR	0.0024
2	Material	VMC	0.0068
3	Capital goods, service	DE	0.017
4	Telecommunication	CTL	0.029
5	Living essential	CPB	0.043
240	Living essential	ADM	16.15
241	Living essential	KO	19.25
242	Consumer goods, service	F	19.63
243	Living essential	PEP	24.14
244	Telecommunication	T	30.91

Furthermore, the daily average of stock prices of the top five randomness is represented by H, as well as the corresponding quantity of the bottom five randomness is represented by L; in the graph it is computed by

$$E_k = \frac{1}{5}\sum_{i=1}^{5} S(i, k) \tag{5.2}$$

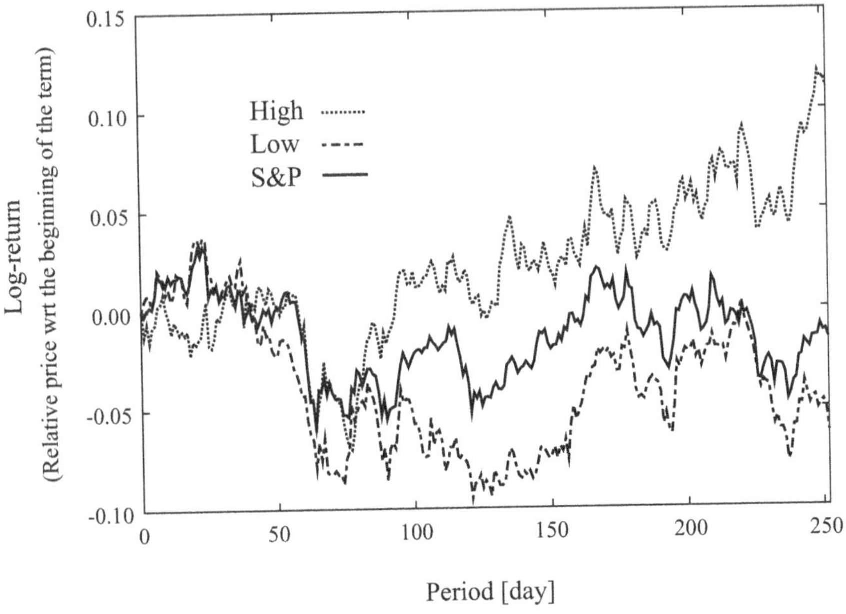

Fig. 5.10 The performance of H/L in 1994 of the five most/least random stocks in 1993

Figures 5.10, 5.11, 5.12, and 5.13 show the three lines of H, L, and S&P500 for 1994, 1995, 1996, and 1997, respectively, where H and L are the average of the top five and the bottom five in terms of the randomness level in the previous year, 1993, 1994, 1995, and 1996, respectively.

Throughout the entire period of 4 years, H is larger than S&P500 and L is smaller than S&P500. However, this ordering does not begin immediately. It takes a few months after the testing period starts after measuring the randomness. Nevertheless, this result supports the rule to some extent, although the inequality

$$\text{Average} < \text{H}$$

is not as strong as the inequality

$$\text{L} < \text{Average}$$

as can be seen in Fig. 5.11 where the line for H almost overlaps the average in most terms.

Combining the two results for the falling market of 2007–2010 in Japan and the rising market of 1993–1997 in the USA, stocks whose price fluctuation (microscopic) is highly random are relatively stable in the following year, and stocks whose price fluctuation is less random are likely to fall in the following year. This is a convincing result for the falling market. However, we do not have a strong

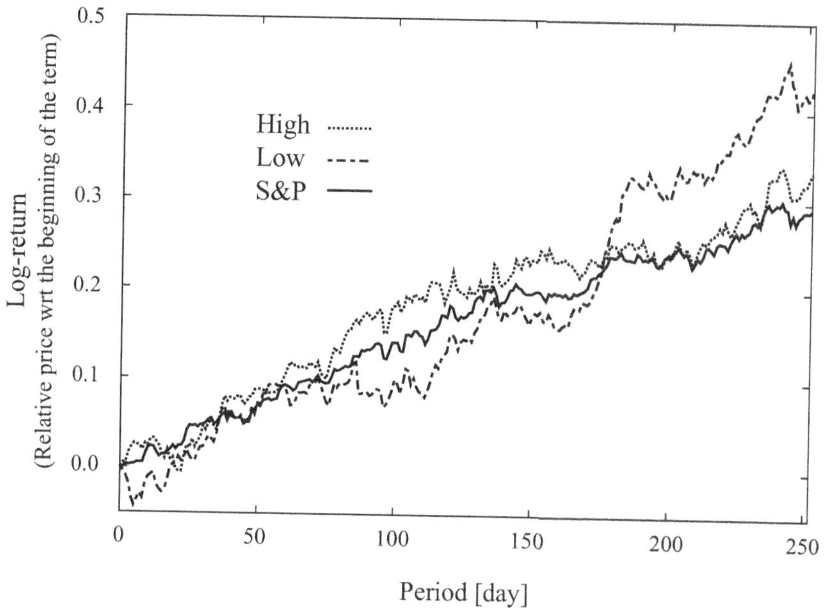

Fig. 5.11 The performance of H/L in 1995 of the five most/least random stocks in 1994

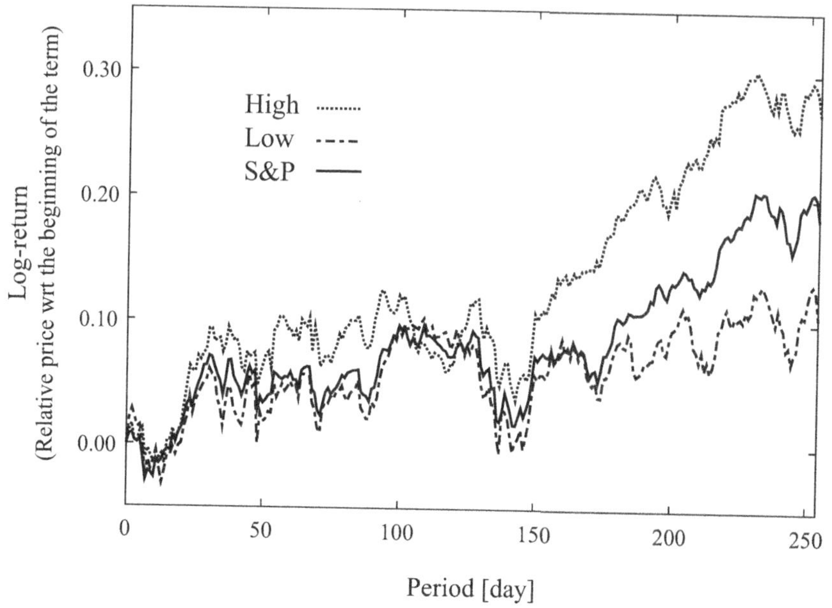

Fig. 5.12 The performance of H/L in 1996 of the five most/least random stocks in 1995

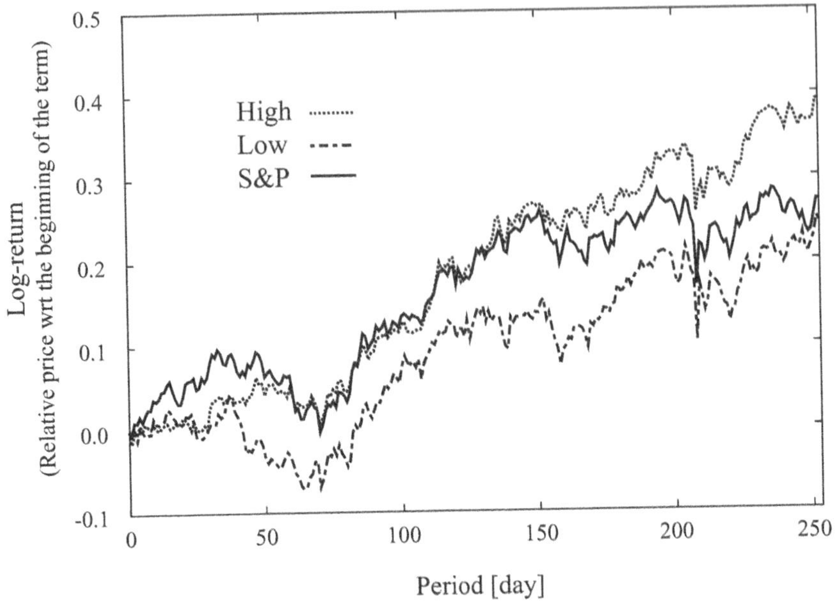

Fig. 5.13 The performance of H/L in 1997 of the five most/least random stocks in 1996

evidence to support the possibility of highly random stocks in context of the RMT-test used on the 1-minute resolution data of price fluctuation performing better in the following year compared to less random stocks. Occasionally the opposite examples tend to occur where the less random stocks surge over the market average. This is a reasonable conclusion considering that fear of losing money is much stronger than the pleasure of gaining money. More to say, it is harder to predict the rise of a stock price than to predict the fall of a stock price.

Since we had limited data at the time when we conducted this experiment, we tried using the NYSE tick data from 1993 to 2002, which we used in the first study in Chap. 3, with less-than-ideal results. That is, although the period 1993–1996 is a period of rising stock prices, it was not possible to generate 1-minute data because the trading density was still small during this period, and we were limited to selecting hourly representative values. The results showed that there was no strong correlation between the order of randomness and the rise and fall of stock prices in the following year. The data intervals are becoming more closely spaced, partly due to the increase in online trading, and we believe that using more recent data would provide some insight.

Compared to the five stocks with the lowest randomness, many of the stocks with the highest randomness are very well known and appear multiple times throughout the 4-year period. In particular, KO, MRK, and WMT are still members of the Dow Jones Industrial Average, while T, PEP, F, and GE were the mainstays of the NYSE in the 1990s and were highly liquid stocks. Considering this, it seems that there may be circumstances that are not revealed by simply calculating the randomness of the S&P500's constituent stocks uniformly on a 1-second or 1-minute basis. The stocks belonging to the lowest randomness members in the analysis here have by far the largest number of trades and are likely to have inherent time intervals that are an order of magnitude shorter than those of the other stocks. Furthermore, the fact that stocks with low randomness in the 1-minute analysis tend to have unstable stock prices in the following year suggests that they are not only prone to fall but also to soar, while stocks with high randomness and stable stock prices are those with low liquidity and prices that do not change much.

5.3 Randomness as a Market Indicator

This section examines the possibility of using the degree of randomness measured by the RMT-test to predict market declines. As a rule, investors who participate in the stock market pursue profit. That is, they buy when the stock price seems low and try to sell when the price is significantly higher than the original price. However, this is a macro price story. Actual prices rise and fall in small increments and are constantly fluctuating. Until the twentieth century, "weekly" and "daily" were the norm. As personal computers equipped with Windows 2000 started to have communication skill, "minutes" and "seconds" came to be used as well. Thus when talking about prices, the resolution needs to be fixed.

Micro prices are constantly fluctuating, rising, and falling in small increments. In Sect. 5.2, we observed the relationship between macro price ups and downs and micro randomness. We found that the lower the micro randomness in a given year with respect to the relative strength of individual stocks, the lower the macro performance in the following year. Moreover, this rule seems to hold in both rising and falling markets. This gives us a hint to prepare for possible market declines by choosing stocks that are relatively safe. Roughly speaking, it is safer to pick stocks with high randomness. However, it was not clear whether high randomness is indicative of future growth potential.

In this section, we will attempt to apply this finding to predict possible market crashes. If we could apply the result that stocks with high randomness have a lower rate of decline in the following year than stocks with low randomness to data from different time periods for a single stock, we might be able to predict a crash. However, if one tries to use RMT-test to calculate the randomness of individual stock prices, one has to use methods such as block ticks, as described in Chap. 3, because tick data have different execution intervals, and there is room for various factors to enter the data preprocessing stage. In addition, if one tries to have at least one representative value for a given time interval, one must be prepared to spend a very long time, such as every hour, as was done in Chap. 3. Therefore, the authors calculated the randomness of the price data at regular intervals using the TOPIX and TOPIXcore30 indices (TOPIX and TOPIXcore30) with the smallest time intervals (every second) distributed for a fee by the Japan Exchange and examined the relationship between the randomness and subsequent price fluctuations. Prior to this study, the authors had conducted a study on the disciplinarity of 1-second intervals for the TOPIX index and individual stock prices in the same market. The background of this study was the controversy surrounding the statistical nature of price fluctuations. Originally, the idea that the mathematical nature of price fluctuations was a random walk had existed since the early twentieth century, but studies using real data on price fluctuations in the late twentieth century began to emphasize aspects of price fluctuations that were not simply random walks, particularly the self-similarity of price fluctuations.

It is necessary here to state an important finding about the nature of price fluctuations. Price fluctuations are unpredictable and have a so-called "random walk" nature. This is true regardless of the resolution. That is, next year's prices are unpredictable, but tomorrow's prices are also unpredictable. Furthermore, it is not even possible to predict exactly whether prices will rise or fall 1 s ahead of time. However, just as it is impossible for the price of a 300-yen apple to be 30 yen tomorrow, or even 3000 yen tomorrow, we must determine the precision of the price as well as the precision of the time interval before we can talk about it. The 1900 Bachelier paper (Bachelier 1900), which is regarded as the starting point for the science of price fluctuations, is significant because it pointed out the "random walking" nature of prices, but later, when price data came to be used as a source of data, the distribution of price fluctuations became not the so-called normal distribution (also called Gaussian distribution), but it is now known that it has a "fat-tail" nature with a wide base. The nature paper by Mantegna and Stanley in 1994 Mantegna and Stanley (1995) drew attention to this, and together with the fractal nature of prices, it has attracted much interest in the field of physical science (Bouchaud and Potters 2000; Mantegna and Stanley 2000). Although a rigorous discussion of the mathematical properties of this distribution function was fashionable for a time, it was not very successful because, as with many fractals found in nature, it is unreasonable to try to generalize a story that is only valid within a certain range of resolution. Furthermore, fat tails also have parameters that depend on the extent of their base, and even if they ever have a fixed value, it should only be within the range of their resolution. The authors investigated this using high-frequency price data. They found the following three points about the nature of high-frequency price fluctuations, even within the range of current trading speed.

(1) Just as the time resolution of data, i.e., annual, monthly, weekly, and daily price fluctuations are different, the distribution function of high-frequency data differs depending on whether the data is for 1-minute or 1-second time frames. As the time resolution of the fluctuation of a single stock price is increased, some of them appear to gradually approach a normal distribution (Tanaka-Yamawaki and Yamanaka 2019).

(2) When compared at the same resolution, the type of distribution function changes depending on the time period, and may be close to the normal distribution, or may deviate from it significantly. See Tanaka-Yamawaki and Ikura (2020).

(3) Price fluctuations of single stock prices are different from price fluctuations of stock indices, which are averages of multiple stock prices; even when single price fluctuations approach the normal distribution, stock index fluctuations do not (Tanaka-Yamawaki and Ikura 2020).

To obtain these results, the authors studied the statistical properties of price fluctuations at different time resolutions, using milliseconds of stock price data and 1-minute index data from the TSE Arrowhead market. As a result, they found that the distribution of the 1-minute index data deviated from a Gaussian distribution and had a wide base, but gradually became closer to a Gaussian distribution as the time resolution was increased. Of course, further research is needed to verify the results, as they did not deal with a large number of data, but it is at least certain that the statistical distribution changes with different resolutions.

Historically, stock price fluctuations were assumed to be random walks (Bachelier 1900). At that time, price fluctuations meant "weekly" or "daily" fluctuations. A random walk occurs when, microscopically speaking, the probability of going to the right is approximately equal to the probability of going to the left. When this is repeated over a long period of time, the probability of eventually staying at x follows a Gaussian (or normal) distribution. Chandrasekhar showed this using a simple formula to derive. This argument is so simple that the author had a very strong impression of it when he was young and recommends that readers read it if they are interested (Chandrasekhar 1943).

By the way, in this chapter, we will use actual data to investigate whether a sudden change in the randomness of the second-order price fluctuation triggers a decline in market prices. As you can see from the results, the forecasting accuracy is not that high. In particular, market price declines can be caused by a variety of factors. Here, we distinguish between declines caused by external factors unrelated to market sentiment as "exogenous" and those caused by worsening market sentiment as "endogenous." The main question is whether the latter can be predicted by looking at sudden changes in the randomness of the second-hand price fluctuations. As the prospect theory (Kahneman and versky 1979) indicates, human emotions are governed by the law that the fear of loss is much stronger than the pleasure of gain. This effect is frequently observed in the stock market.

Black Monday was one of the few events experienced by the author. Black Monday was the stock market crash that occurred on the New York Stock Exchange on Monday, October 19, 1987. At the time, the US market was extremely overheated as Reaganomics drove stock prices higher and everyone began to invest in stocks. The Dow Jones Industrial Average was over \$2200, and people were investing like a game of chicken because they were afraid that it could fall at any time. People's fears had been smoldering for about a year, so it was no surprise when the market crashed at the slightest opportunity. Then, on Monday, October 19, 1987, the crash really happened. On that day, the Dow Jones Industrial Average fell more than \$500 from the previous week. But what caused the decline, the story is not so simple. But the answer to what caused the decline was not so clear. Since then, there have been a number of similar small-scale declines, likely caused by heightened investor anxiety.

It was a somewhat similar situation in 2020–2021, when stock prices turned higher today after a brief decline in the stock market shortly after the Abenomics stock market began its COVID-19 disaster. Investor sentiment has been shaky since the summer of 2021 because the stock market, which by all accounts had risen too much, could fall at any moment. Even the slightest movement would cause stock prices to fall and then rise again.

Naturally, the idea of using the RMT test for Japanese stock indices was born. Of course, not all declines are due to fluctuations in investor sentiment, and stock prices fluctuate all the time, so it is impossible to say what percentage of a decline we would call a crash. However, since Black Monday was also ultimately unknowable and had such a large social impact, various systems were created to prevent the same thing from happening again. A typical example is the "market circuit breaker," which temporarily halts trading when stock prices fluctuate above a certain upper limit.

A preliminary study using TOPIX from 2010 to 2021 (JPX Cloud 2022) found that the parameter $|D_6|$, which indicates deviation from the maximum randomness in the "normal" state, increases from 2010 to 2013 in the first period. The first peak was of the exogenous type due to the external effects of the earthquake and the accident at the Fukushima Daiichi Nuclear Power Plant. Such peaks cannot be predicted because they occur at the same time that prices fall due to shocks from the accident. On the other hand, the last two of the three peaks between 2010 and 2013 belong to the type where prices fall sharply due to instability in investor sentiment. According to prospect theory (Kahneman and versky 1979), investors' fear of losing money is much stronger than their expectation of gaining. In fact, a price decline was observed at the end of March 2012 after the peak of the parameter $|D_6|$ that appeared in November 2011. Details can be found in a recent paper by the authors (Tanaka-Yamawaki and Ikura 2022).

Cross-checking with the randomness index $|D_6|$, $\Delta\lambda = \lambda_1 - \lambda_+$ in Eq. (4.4) is calculated to be $\Delta\lambda = \lambda_1 - 2.488$ for $Q = 3$. For this reason, we created a new program to process the price data (Appendix C). The program is outlined as follows.

(1) The data for each year is divided into pieces of length NL so that $L = QN$.
(2) Each piece is further divided into smaller pieces of length L to construct an $N \times L$ rectangular matrix G, from which the correlation matrix $C = GG^t$ is constructed.
(3) Diagonalize C to obtain N eigenvalues (using Jacobi's rotation algorithm).
(4) Compute $\Delta\lambda = \lambda_1 - \lambda_+$ as well as D_k for $k = 2, 3, \ldots, 6$ and output them.
(5) Plot a histogram of the eigenvalue distribution.

⟨Case study: TOPIX Index October 2010–January 2022⟩

The input data consist of 11 sets of second-by-second historical tick data for TOPIX from October 2010 to January 2022. Each set consists of price time series for each year. For example, each set out of the first ten sets has price time series

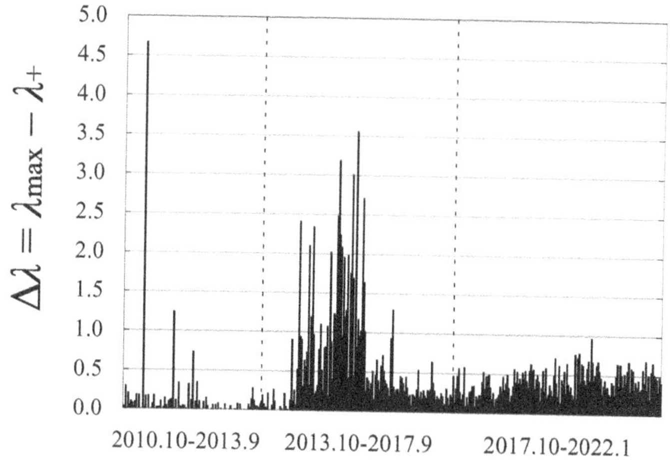

Fig. 5.14 Deviation from randomness measured by $\Delta\lambda$ in the 1-second price fluctuation of TOPIX index

from the first business day of October to the last business day of September in the next year, except the 11th set that starts from the first business day of October 2021 to the last business day of January 2022. The output of this program is the following three items.

(1) $\Delta\lambda = \lambda_1 - \lambda_+$, where λ_1 is the maximum eigenvalue and λ_+ is the upper bound of the RMT spectrum.
(2) Set of D_k, the deviation from the random limit, representing the inverse of randomness.
(3) Histogram of the eigenvalue distribution.

The output result for $\Delta\lambda$ is plotted in Fig. 5.14 for the entire period of October 2011 to January 2022. Based on the behavior of $\Delta\lambda$, the entire period October 2010–January 2022 can be divided into three parts: Part 1, October 2010–September 2013, is characterized by the smallness of $\Delta\lambda$, except the three large peaks, indicating that the price fluctuations at 1-second resolution are close to the random walk in this period. On the other hand, Part 2, October 2013–September 2017, has full of large amplitude in the deviation from random limit $\Delta\lambda$ that floats from zero throughout the period. This tendency of non-random walk behavior in $\Delta\lambda$ continues throughout Part 3, October 2017–January 2022, but the heights of the peaks are all smaller than Part 2.

In Part 1, there are three distinct peaks, as shown in Fig. 5.15. The first and the biggest peak is in the term March 7–17. This peak is very abrupt and occurred without any preliminary sign. As shown in the leftmost table of Table 5.13 (i), the daily price fell simultaneously in this term. Note that there is no particular sign in the previous term nor the next term. The reason is clear. This is caused by an

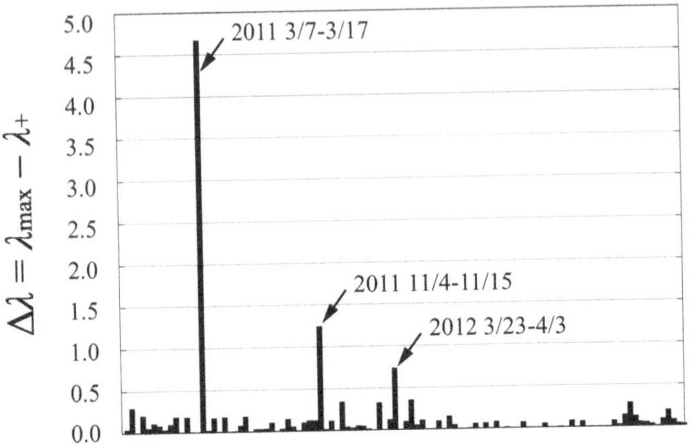

Fig. 5.15 Three notable peaks of $\Delta\lambda$ are observed in Part 1 period of October 2011–December 2013

Table 5.13 Three peaks in Fig. 5.15 consist of two types, exogenous in (i) and endogenous in (ii) and (iii)

(i)

Term	$\Delta\lambda$	TOPIX index
2011 2/24-3/7	0.00	-
2011 3/7 -3/17	4.67	Decline
2011 3/17-3/29	0.02	-

(ii)

Term	$\Delta\lambda$	TOPIX index
2011 10/24-11/4	0.13	-
2011 11/4 -11/15	1.24	-
2011 11/15-11/25	0.02	Decline
2011 11/25-12/6	0.12	-

(iii)

Term	$\Delta\lambda$	TOPIX index
2012 3/13-3/23	0.12	-
2012 3/23-4/3	0.73	-
2012 4/3 -4/11	0.00	Decline
2012 4/11-4/20	0.10	-

exogenous reason, the 3.11 East Japan Earthquake and the accident in the nuclear plant, Fukushima 1, operated by Tokyo Electric Power Supply.

On the contrary, the rest of two peaks found in the terms from November 4, 2011, to November 15, 2011, and from March 23, 2012, to April 3, 2012, are not as large as the first peak, but the daily price falls in the next term or in the vicinity of the next term. This can be assumed by the internal reason. We can call this as the endogenous-type price fall or crash. Note there is a sudden increase of $\Delta\lambda$ in the term before a price fall occurs, as shown in Table 5.13 (ii) as well as (iii).

More detailed description of those three peaks is given in Fig. 5.16, where the upper three figures show the movements of daily prices and the lower three figures show the three outputs, 1), 2), and 3), stated above from the program (given in Appendix C). In (i), the actual price fall and its recovery in the term of March 7-March 11 are plotted in the upper figure, corresponding to the increase of $\Delta\lambda$ from nearly zero to $\Delta\lambda = 4.76$ in this term and immediate return to 0.02 in the next term. The lower figure at the leftmost side shows the output 3) from which we can read out $\Delta\lambda = 4.76$ as the departure of the largest eigenvalue located at around $\lambda_1 = 7.2$ from the theoretical upper limit of $\lambda_+ = 2.488$.

Likewise, the upper figure in (ii) shows the actual price fall of the daily price of TOPIX index in the term of November 16–November 25, 2011, resulting from the

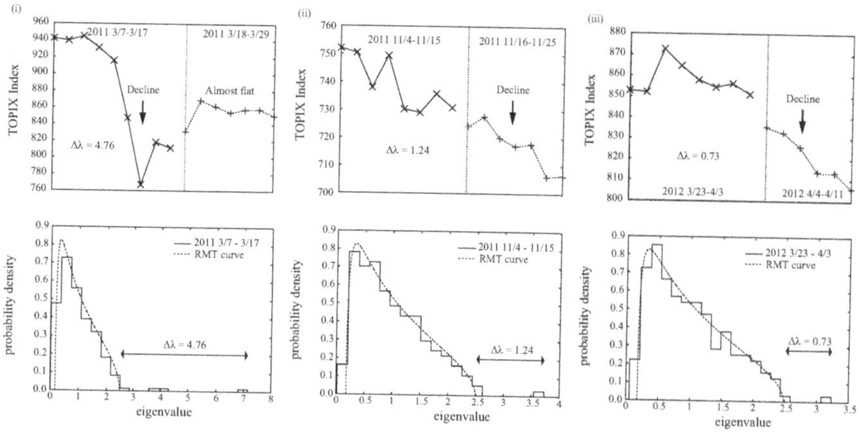

Fig. 5.16 Peaks of $\Delta\lambda$ in the lower graphs of (i)–(iii) are compared to the daily prices in each term, followed by subsequent declines in Table 5.13

sudden deterioration of randomness shown in the lower graph, the output 3) from which we can read out $\Delta\lambda = 1.24$ for the term November 4–November 15, 2011 as the departure of the largest eigenvalue located at around $\lambda_1 = 3.72$ from the theoretical upper limit at $\lambda_+ = 2.488$.

Similarly, the upper figure in (iii) shows the actual price fall of the daily price of TOPIX index in the term of April 4–April 11, 2012, resulting from the sudden deterioration of randomness shown in the lower graph, the output 3) from which we can read out $\Delta\lambda = 0.73$ for the term March 23–April 3, 2012, as the departure of the largest eigenvalue located at around $\lambda_1 = 3.21$ from the theoretical upper limit at $\lambda_+ = 2.488$.

Incidentally, the program outputs a set of $|D_k|$ together with $\Delta\lambda$. For our purpose, however, $\Delta\lambda$ seems sufficient. From now on, we stick to analyze the behavior of $\Delta\lambda$ and use $|D_k|$ only in the case of checking consistency.

⟨Case study: TOPIXcore30 Index January 2011–January 2022⟩

The analysis of TOPIX shows that a sudden increase in $\Delta\lambda$ triggers a decline in the daily chart when the 1-second fluctuations are in the normal state of a random walk. However, this analysis does not work because $\Delta\lambda$ remains large after 2013. Therefore, in addition to TOPIX, for which 1-second fluctuation data exists for a long period of time, we conducted an analysis using another indicator, the Nikkei 400 and TOPIXcore30. However, since the Nikkei 400 also has a large $\Delta\lambda$ after 2013, and the nature of random walks is not apparent, TOPIXcore30 was chosen as the next target.

The TOPIXcore30 index is a weighted average of 30 indexes of selected companies listed on the First Section of the Tokyo Stock Exchange that are highly liquid and have large market capitalization. This index is also published by the Tokyo Stock Exchange on a second-by-second basis (JPX cloud, 2022). The results

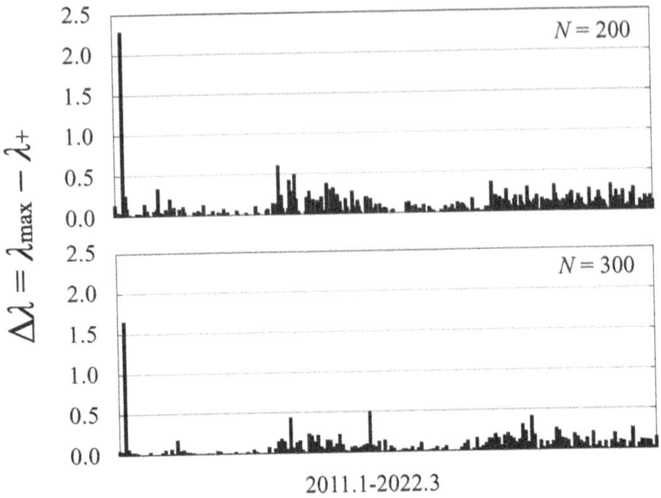

Fig. 5.17 The $\Delta\lambda$ for TOPIXcore30 index in January 2011–March 2022 is plotted for $N = 200$ and $N = 300$

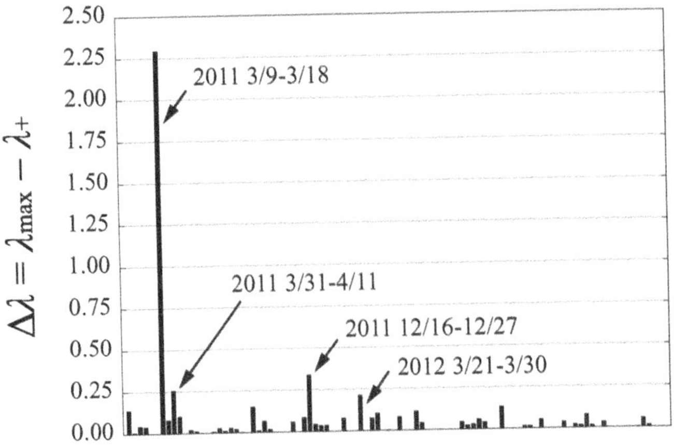

Fig. 5.18 The $\Delta\lambda$ for TOPIXcore30 index in the first period of January 2011–December 2013 is shown

are shown in Fig. 5.17 for $N = 200$ and $N = 300$. As expected, the value of $\Delta\lambda$ is smaller in the results for $N = 300$.

Unlike TOPIX and Nikkei 400, TOPIXcore30 observes $\Delta\lambda \simeq 0$ throughout the entire period, except for some non-zero peaks, indicating that the price movement is more similar to a random walk. This can be understood as the random walk nature of the 30 highly liquid stocks in TOPIXcore30 being maintained under normal circumstances. Figure 5.18 shows the first part of Fig. 5.17 with $N = 200$.

First, we check the consistency with the TOPIX data. In Figs. 5.18 and 5.15, the March 2011 peak and the March 2012 peak are observed. However, the November 2011 peak observed for TOPIX in Fig. 5.15 is not observed in Fig. 5.18, and a new peak appears in December 2011. Thus, we can see that TOPIX and TOPIXcore30 have been moving almost identically before October 2013. As a result, we may conclude that endogenous price declines are predictable due to the sharp deterioration in randomness as indicated by the second and third peaks, while exogenous price declines are not signaling in advance.

The remainder of the post-2013 period does not show a clear peak against the backdrop of a random walk. The major problem with this failure stems from the fact that the basic assumption of a random walk, indicated by $\Delta\lambda \simeq 0$ in normal times, is not observed in this region. This feature is also observed in the Nikkei 400 index, but its deviation from the random walk is much more gradual than that of TOPIX. The reason for this can be attributed to the difference in the size of firms in the two indices, with more than 2000 firms in the TOPIX and 400 firms in the Nikkei 400; given the changes in the TSE system between 2013 and 2015, many smaller firms in the TOPIX would have been significantly affected.

Therefore, instead of TOPIX, we analyze the TOPIXcore30. This index is a weighted average of 30 stocks selected from firms listed on the First Section of the Tokyo Stock Exchange that have high liquidity and large market capitalization. This index is also published by the Tokyo Stock Exchange on a second-by-second basis. Unlike TOPIX and Nikkei 400, $\Delta\lambda \simeq 0$ is observed for TOPIXcore30 throughout the entire period except for some non-zero peaks, indicating that the price movement is more similar to a random walk. This can be understood as the random walk nature of the 30 most liquid stocks in the TOPIXcore30 being maintained under normal conditions.

From now on, we consider some significant peaks appeared in the case of TOPIXcore30, in the period of January 6, 2014–March 11, 2022, for $N = 300$. We shall focus on the peaks $\Delta\lambda \geq 0.25$. There are six such peaks found in Fig. 5.19. They are numbered (i)–(vi) in the chronological order. Let us look at them one by one.

Although there are six peaks exceeding 0.25, not all of them are in the environment of random walk. In particular, the peak (iii) is too close to the next peak (iv), so that there is no room between (iii) and (iv) to return to the random walk environment. Similar situation holds for the peak (iv) as can be seen in Fig. 5.19. This may be a part of the reason why the clear price decline is not recognized after those peaks in Fig. 5.20 and Table 5.14. On the contrary, there are sufficiently large rooms after the peaks (i), (ii), (v), and (vi) as can be recognized in Fig. 5.19. This can be understood for the reason that the price declines are well recognized after them.

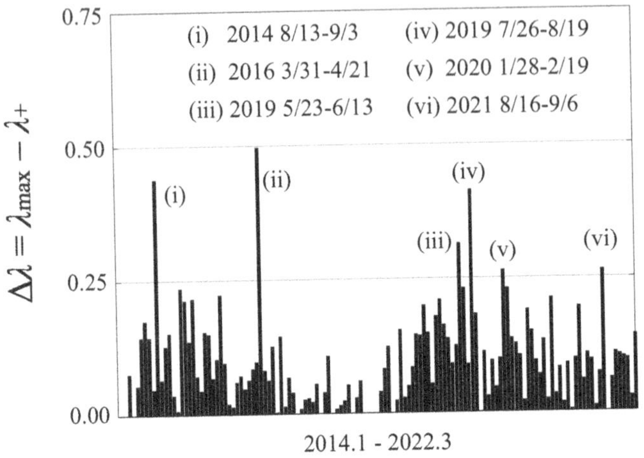

Fig. 5.19 There are six peaks exceeding 0.25, relatively larger than other peaks

Throughout this section, we have discussed the possibility of predicting a market crash by using the RMT-test. The parameter $\Delta\lambda$ was used to represent the degree of randomness of the second-by-second price fluctuation, where $\Delta\lambda$ is zero for a random sequence and is large if not random. We have searched for evidences of sudden increase of $\Delta\lambda$ to predict a future crash, in the time series of two stock indices, TOPIX and TOPIXcore30, in the period of October 2010–January 2022. It was quite successful in the beginning part of the data from October 2010 to the middle of 2013, because the basic behavior of fluctuation was random walk in this period. There, $\Delta\lambda$ is zero except three or four sharp peaks. Although the first peak in March 2011 was the exogenous type caused by 3.11 Earthquake and the nuclear power plant accident, the rest of the peaks in March 2012 and others are the endogenous type and triggered market declines afterward.

On the other hand, the later period after mid-2013 to January 2022 was difficult. The random nature is weak in TOPIXcore30, and not observed in TOPIX index. Nevertheless, we picked up six vague peaks in TOPIXcore30 and examined if any of them trigger the market decline. We found four positive examples out of six candidates. The above conclusion is based on somewhat limited information, from the analysis of TOPIX and TOPIXcore30 in the period of October 2010–January 2022, for $N = 200$ and $N = 300$. In the future, we would like to collate the results with those of other data and analyses using different parameters such as N larger than 300.

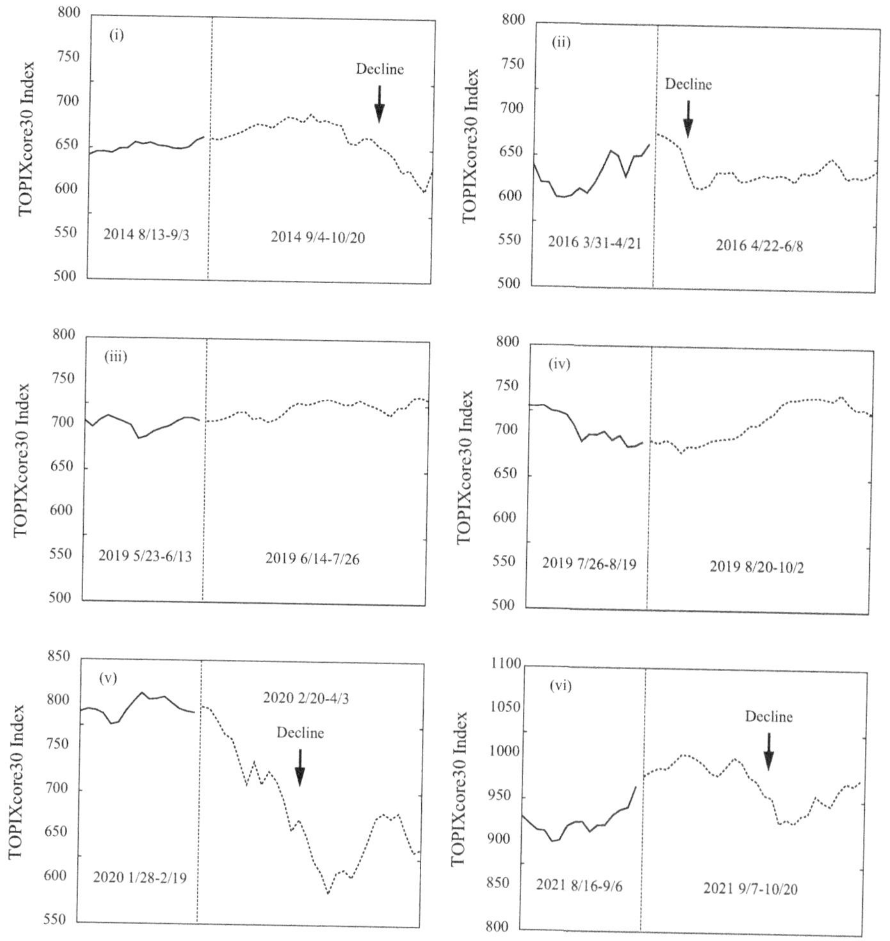

Fig. 5.20 Price tends to decline after relatively large peaks appear

Table 5.14 Price tends to decline after $\Delta\lambda$ increases

(i)

Term	$\Delta\lambda$	TOPIXcore30 index
2014 8/3 -8/13	0.05	-
2014 8/13-9/3	0.44	-
2014 9/3 -9/26	0.07	-
2014 9/26-10/20	0.13	Sharp Decline

(ii)

Term	$\Delta\lambda$	TOPIXcore30 index
2016 3/9 -3/31	0.10	-
2016 3/31-4/21	0.50	-
2014 4/21-5/18	0.08	Sharp Decline
2014 5/18 -6/8	0.06	-

(iii)

Term	$\Delta\lambda$	TOPIXcore30 index
2019 4/24-5/23	0.13	-
2019 5/23-6/13	0.32	-
2019 6/13-7/4	0.23	-
2019 7/4 -7/26	0.09	-

(iv)

Term	$\Delta\lambda$	TOPIXcore30 index
2019 7/4 -7/26	0.09	-
2019 7/26-8/19	0.41	-
2019 8/19-9/9	0.18	-
2019 9/9 -10/2	- 0.01	-

(v)

Term	$\Delta\lambda$	TOPIXcore30 index
2020 1/6 -1/28	0.10	-
2020 1/28-2/19	0.26	-
2020 2/19-3/12	0.23	Sharp Decline
2020 3/12-4/3	0.14	Sharp Decline

(vi)

Term	$\Delta\lambda$	TOPIXcore30 index
2021 7/21-8/16	0.07	-
2021 8/16-9/6	0.26	-
2021 9/6 -9/29	-0.03	-
2021 9/29-10/20	0.06	Sharp Decline

Chapter 6
Conclusion

We have explored the world of numerical research armed with RMT-oriented methodologies for big data analysis. At first glance, the methodology may appear to be very mathematical. However, in the course of our research, we have learned that once the process is fully formulated and the algorithms are established, it should be a tool that anyone can use. That is how we enjoyed working together and struggling with a huge size of data sets. Such simplicity is at the heart of a technology suitable for big data analysis. The authors strongly hope that the two methodologies introduced here, RMT-PCA and RMT-test, will inspire a wider range of applications and many ideas that will emerge in the future for our readers.

The first methodology RMT-PCA is designed to extract the principal components from a set of N different data of the equal time data of the length L, where N and L are large. The methodology is powerful and simple, as far as the data in the required structure is prepared. Yes, the difficulty is how to prepare the data. Stock prices seem to be an ideal example. However, transactions do not always take place at the same moment. We attempted to apply RMT-PCA on three different data under very limited on data acquisition circumstances. Nevertheless, we succeeded in extracting historical changes in active sectors from banks (code numbers in 83XX) to public services (codes numbers in 95XX) between 2007 and 2009. We also succeeded in detecting the sub-prime mortgage-related panic in the third quarter of 2007, the Lehman shock in the fourth quarter of 2008, and its recovery of the market, using only numerical calculations of high-frequency price data.

The second method, RMT-test, aims to measure the degree of randomness of very long data sequences. As we have seen in this book, if the degree of randomness can serve as a signal, this method would have a wide range of applications.

The basis of the method is the eigenvalue distribution of the correlation matrix, similar to that used in the first method, RMT-PCA. If N is infinitely large while keeping the parameter Q fixed, which is the condition for RMT, the randomness is high if it matches the Marcenko-Pastur function, which depends only on Q, and low if it does not. The purpose of the RMT-test is to measure the degree of

M. Tanaka-Yamawaki, Y. Ikura, *Principal Component Analysis and Randomness Test for Big Data Analysis*, Evolutionary Economics and Social Complexity Science 25, https://doi.org/10.1007/978-981-19-3967-9_6

randomness by quantifying the degree of agreement. Using the fact that the order k of the moment D_k is a free parameter, we chose $k = 6$ in order to have sufficient precision to distinguish the difference between two well-known machine random numbers, linear congruence generator (LCG) and Mersenne Twister (MT). We have also found that the same $k = 6$ is sufficient to compare the three physical random numbers (PRNs), which are considered superior in terms of non-reproducibility. Based on this process, we established $|D_6|$ defined in Eq. (4.33) as a main tool to measure the degree of randomness.

Furthermore, it was found that the three physical random numbers (PRNs), which were considered superior to the two in terms of non-reproducibility, require order $k = 6$ to be accurate enough to compare them, and to compare the randomness of such high randomness data, the deviation between the 6th-order moment μ_k and the corresponding RMT theoretical value μ_k can be used to compare the degree of randomness of such high randomness data.

By the way, the purpose of the RMT-test is not to distinguish randomness or not but to measure the degree of randomness for a given sequence. For this purpose, it is necessary to prepare data with a very low randomness, but it is difficult to find such data in the natural world. Therefore, recalling the well-known example of discarding 500 initial data of LCGs because of their low randomness, we first set various initial values (SEED) and created an artificial data set consisting of only the initial 500 LCGs concatenated together. Also, we artificially lowered randomness by converting the output of LCG and MT to the log-return sequences.

At last, we created artificial data by sequentially shuffling the regular data and compared the randomness measured by the RMT test using this data with the widely used NIST test. The data were then compared with the widely used NIST test. The randomness of the low randomness data thus created is extremely large when using the above-mentioned $|D_6|$, but by comparing the values, it is possible to compare the randomness between the data. The comparison of the old and new Hash functions showed that the output of the newer SHA-1 was clearly more random than that of the older MD5.

In Sect. 5.2 of Chap. 5, we considered using this for stock selection, and we used the 1-minute TOPIX 500 stock price data for 2007–2009 used in Chap. 3 to measure the randomness from 1 year of price data for each stock to verify whether there is a relationship between the randomness and performance. The results are reported. The results showed that for most periods except the period of the Lehman Shock, the higher the randomness, the safer the stock. A safe stock is a stock that declines slowly. Indeed, many of the stable electric power stocks (95XX) had high randomness. The randomness of auto stocks (72XX), which fell sharply during this period, tends to be low. However, this is during a period when stock prices tended to fall as a whole, so data from a different period is needed to see how they fare in an up market.

In Sect. 5.3 we conducted a survey of stock price indices, assuming that the degree of randomness of high-frequency price fluctuations in stock indices can serve as a signal of instability in trader psychology. To this end, we introduced a simple indicator $\Delta\lambda = \lambda_1 - \lambda_+$, where λ_1 is the maximum eigenvalue and λ_+ is the upper bound of the RMT spectrum. This is a measure of deviation from the random walk. We found that after the time series of the random walk is strongly disturbed by some fear, $\Delta\lambda$ suddenly changes from zero to a large value (creating a peak in the background of the random walk); then it works to predict a market crash or some decline.

This new indicator $\Delta\lambda$ seems to work in many cases to predict market declines, including TOPIXcore30 in the broad period of 2014–2022 and TOPIX index in the period of 2010–2013, where the price follows the random walk, shown by $\Delta\lambda \simeq 0$ in most times except a few exceptional occasions of sudden increase of $\Delta\lambda$ as the single peaks. Several such examples have been identified in the 2010–2013 TOPIX and 2014–2022 TOPIXcore30. While these are still a small fraction of the evidence to be shown, we believe that the power of analyzing large amounts of numerical data has been proven to some extent by RMT methodologies.

Appendix A
Introduction to Vector, Inner Product, Correlation Matrix

It is sometimes helpful to imagine a concrete image of some famous companies for $i = 1, 2, 3, \ldots, N$. For example, if $i = 1$ is Microsoft and its return is 0.1 at the first moment and -0.2 at the next moment, etc. and if $i = 2$ for Apple Computer and its return is 0.3 at the first moment and -0.1 at the next moment, etc., then the matrix G would be

$$G = \begin{pmatrix} 0.1 & -0.2 & \cdots & g_{1,L} \\ 0.3 & -0.1 & \cdots & g_{2,L} \\ \vdots & \vdots & \ddots & \vdots \\ g_{N,1} & g_{N,2} & \cdots & g_{N,L} \end{pmatrix} = \begin{pmatrix} \text{Msoft}_1 & \text{Msoft}_2 & \cdots & \text{Msoft}_L \\ \text{Apple}_1 & \text{Apple}_2 & \cdots & \text{Apple}_L \\ \vdots & \vdots & \ddots & \vdots \\ \text{stockN}_1 & \text{stockN}_2 & \cdots & \text{stockN}_L \end{pmatrix}$$

Before proceeding further, we recall the definition of the inner product. The inner product can be defined for two vectors of the same dimension, such as $\boldsymbol{g}_i = (g_{i,1}, \cdots, g_{i,L})$ and $\boldsymbol{g}_j = (g_{j,1}, \cdots, g_{j,L})$, and is defined by the sum of the products of two elements at each dimension. Namely, take the first element $g_{i,1}$ from the vector $\boldsymbol{g}_i$ and take $g_{j,1}$ from the vector $\boldsymbol{g}_j$ and multiply them, and doing the same for all the L dimensions then make a sum.

$$\boldsymbol{g}_i \cdot \boldsymbol{g}_j = g_{i,1}g_{j,1} + g_{i,2}g_{j,2} + \cdots + g_{i,L}g_{j,L}$$

In the case of $L = 2$, take $\boldsymbol{x} = (a, b)$ and $\boldsymbol{y} = (c, d)$, and the inner product of $\boldsymbol{x}$ and $\boldsymbol{y}$ is

$$\boldsymbol{x} \cdot \boldsymbol{y} = ac + bd = |\boldsymbol{x}||\boldsymbol{y}| \cos\theta$$

The latter expression is more familiar to us in the high school textbooks. It is clear that the two expressions are identical if we recall that a vector $\boldsymbol{x} = (a, b) = (|\boldsymbol{x}| \cos\alpha, |\boldsymbol{x}| \sin\alpha)$, by means of its magnitude $|\boldsymbol{x}|$ and angle to the x-axis α, and

M. Tanaka-Yamawaki, Y. Ikura, *Principal Component Analysis and Randomness Test for Big Data Analysis*, Evolutionary Economics and Social Complexity Science 25, https://doi.org/10.1007/978-981-19-3967-9

likewise $\boldsymbol{y} = (c, d) = (|\boldsymbol{y}| \cos \beta, |\boldsymbol{y}| \sin \beta)$, by means of its magnitude $|\boldsymbol{y}|$ and angle to the x-axis β. Combining all those, $\boldsymbol{x} \cdot \boldsymbol{y} = ac + bd = |\boldsymbol{x}||\boldsymbol{y}| \cos(\alpha - \beta)$. This means that the value of inner product shows the degree of overlap between two vectors $\boldsymbol{x}$ and $\boldsymbol{y}$.

Appendix B
Jacobi's Rotation Algorithm and Program for the RMT-PCA

Jacobi's rotation algorithm to obtain eigenvalues of a symmetric matrix is as follows:

(1) Find the largest (in absolute value) off-diagonal element $c_{k,l}$ such that $|c_{k,l}| = \max\{c_{i,j}\}$.
(2) Apply the rotation matrix $W(\theta)$ to the k-th row and the l-th column, such that
$W_{k,k} = \cos\theta = W_{l,l}$ and $W_{k,l} = -\sin\theta = -W_{l,k}$,
$W_{i,i} = 1$ for $i \neq k, l$,
$W_{i,j} = 0$ for $i \neq j$, and $i \neq k, l$
such that

$$W(\theta) = \begin{pmatrix} 1\,0 & & & & & & 0\,0 \\ 0\,1 & & & & & & 0\,0 \\ & \ddots & & & & & \\ & & \cos\theta & 0 \cdots 0 & -\sin\theta & & \\ & & 0 & 1 \quad & 0 & & \\ & & \vdots & \ddots & \vdots & & \\ & & 0 & \quad 1 & 0 & & \\ & & \sin\theta & 0 \cdots 0 & \cos\theta & & \\ & & & & & \ddots & \\ 0\,0 & & & & & & 1\,0 \\ 0\,0 & & & & & & 0\,1 \end{pmatrix} \tag{B.1}$$

M. Tanaka-Yamawaki, Y. Ikura, *Principal Component Analysis and Randomness Test for Big Data Analysis*, Evolutionary Economics and Social Complexity Science 25, https://doi.org/10.1007/978-981-19-3967-9

(3) Rotate C by W such that

$$\begin{aligned} W^{-1}CW &= \begin{pmatrix} \cos\theta & \sin\theta \\ -\sin\theta & \cos\theta \end{pmatrix} \begin{pmatrix} c_{i,i} & c_{i,j} \\ c_{i,j} & c_{j,j} \end{pmatrix} \begin{pmatrix} \cos\theta & -\sin\theta \\ \sin\theta & \cos\theta \end{pmatrix} \\ &= \begin{pmatrix} \cos\theta & \sin\theta \\ -\sin\theta & \cos\theta \end{pmatrix} \begin{pmatrix} \cos\theta c_{i,i} + \sin\theta c_{i,j} & \cos\theta c_{i,j} - \sin\theta c_{i,i} \\ \cos\theta c_{i,j} + \sin\theta c_{j,j} & \cos\theta c_{j,j} - \sin\theta c_{i,j} \end{pmatrix} \\ &= \begin{pmatrix} \tilde{c}_{i,i} & 0 \\ 0 & \tilde{c}_{j,j} \end{pmatrix} \end{aligned}$$

⟨Example⟩
For example,

$$C = \begin{pmatrix} 1 & 1/2 & 1/2 \\ 1/2 & 1 & 1/2 \\ 1/2 & 1/2 & 1 \end{pmatrix}$$

can be diagonalized as follows:
[1st rotation] $C \rightarrow C'$

$$\begin{aligned} C' &= V_{12}^t C V_{12} \\ &= \begin{pmatrix} 1/\sqrt{2} & 1/\sqrt{2} & 0 \\ -1/\sqrt{2} & 1/\sqrt{2} & 0 \\ 0 & 0 & 1 \end{pmatrix} \begin{pmatrix} 1 & 1/2 & 1/2 \\ 1/2 & 1 & 1/2 \\ 1/2 & 1/2 & 1 \end{pmatrix} \begin{pmatrix} 1/\sqrt{2} & -1/\sqrt{2} & 0 \\ 1/\sqrt{2} & 1/\sqrt{2} & 0 \\ 0 & 0 & 1 \end{pmatrix} \\ &= \begin{pmatrix} 3\sqrt{2}/4 & 3\sqrt{2}/4 & \sqrt{2}/2 \\ -\sqrt{2}/4 & \sqrt{2}/4 & 0 \\ 1/2 & 1/2 & 1 \end{pmatrix} \begin{pmatrix} 1/\sqrt{2} & -1/\sqrt{2} & 0 \\ 1/\sqrt{2} & 1/\sqrt{2} & 0 \\ 0 & 0 & 1 \end{pmatrix} \\ &= \begin{pmatrix} 3/2 & 0 & 1/\sqrt{2} \\ 0 & 1/2 & 0 \\ 1/\sqrt{2} & 0 & 1 \end{pmatrix} \end{aligned}$$

[2nd rotation] $C' \to C''$

$$
\begin{aligned}
C'' &= V_{13}^t C' V_{13} \\
&= \begin{pmatrix} \cos\theta & 0 & \sin\theta \\ 0 & 1 & 0 \\ -\sin\theta & 0 & \cos\theta \end{pmatrix} \begin{pmatrix} 3/2 & 0 & 1/\sqrt{2} \\ 0 & 1/2 & 0 \\ 1/\sqrt{2} & 0 & 1 \end{pmatrix} \begin{pmatrix} \cos\theta & 0 & -\sin\theta \\ 0 & 1 & 0 \\ \sin\theta & 0 & \cos\theta \end{pmatrix} \\
&= \begin{pmatrix} \frac{3}{2}\cos\theta + \frac{1}{\sqrt{2}}\sin\theta & 0 & \frac{1}{\sqrt{2}}\cos\theta + \sin\theta \\ 0 & \frac{1}{2} & 0 \\ \frac{1}{\sqrt{2}}\cos\theta - \frac{3}{2}\sin\theta & 0 & \cos\theta - \frac{1}{\sqrt{2}}\sin\theta \end{pmatrix} \begin{pmatrix} \cos\theta & 0 & -\sin\theta \\ 0 & 1 & 0 \\ \sin\theta & 0 & \cos\theta \end{pmatrix} \\
&= \begin{pmatrix} \frac{5}{4} + \frac{\sin 2\theta}{\sqrt{2}} + \frac{\cos 2\theta}{4} & 0 & \frac{\cos 2\theta}{\sqrt{2}} - \frac{\sin 2\theta}{4} \\ 0 & \frac{1}{2} & 0 \\ \frac{\cos 2\theta}{\sqrt{2}} - \frac{\sin 2\theta}{4} & 0 & \frac{5}{4} - \frac{\sin 2\theta}{\sqrt{2}} - \frac{\cos 2\theta}{4} \end{pmatrix}
\end{aligned}
$$

where θ satisfies $\tan 2\theta = 2c'_{1,3}/(c'_{1,1} - c'_{3,3}) = 2\sqrt{2}$, and we obtain the following results:

$$
C'' = \begin{pmatrix} 2 & 0 & 0 \\ 0 & 1/2 & 0 \\ 0 & 0 & 1/2 \end{pmatrix}
$$

and diagonalization matrix

$$
\begin{aligned}
V_{12}V_{13} &= \begin{pmatrix} 1/\sqrt{2} & -1/\sqrt{2} & 0 \\ 1/\sqrt{2} & 1/\sqrt{2} & 0 \\ 0 & 0 & 1 \end{pmatrix} \begin{pmatrix} \cos\theta & 0 & -\sin\theta \\ 0 & 1 & 0 \\ \sin\theta & 0 & \cos\theta \end{pmatrix} \\
&= \begin{pmatrix} \cos\theta/\sqrt{2} & -1/\sqrt{2} & -\sin\theta/\sqrt{2} \\ \cos\theta/\sqrt{2} & 1/\sqrt{2} & -\sin\theta/\sqrt{2} \\ \sin\theta & 0 & \cos\theta \end{pmatrix} \\
&= \begin{pmatrix} \sqrt{3}/3 & -\sqrt{2}/2 & -\sqrt{6}/6 \\ \sqrt{3}/3 & \sqrt{2}/2 & -\sqrt{6}/6 \\ \sqrt{3}/3 & 0 & \sqrt{6}/3 \end{pmatrix}
\end{aligned}
$$

⟨PMT-PCA programs⟩

Listing B.1 RNPD.c

```
/* 1 This program computes the RMT-theoretical Distribution curve */
#include<stdio.h>
#include<stdlib.h>
#include<math.h>

#define SEPARATE 10000

//./EXE_fileSIZE TIME
int main(int argc, char *argv[])
{

  int i = 0, l = atoi(argv[2]), n =atoi(argv[1]);
  double q,qq,lam_max,lam_min,lam[SEPARATE + 1],r_lam[SEPARATE + 1];
  double part;
  FILE *fp;
  char fname[20];

  /*printf("Input N:");
  scanf("%d",&n);

  printf("Input L:");
  scanf("%d",&l);
  */

  q = (double)l / n;
  /*
    if(q <=1.0){
    printf(" L < N TERMINATING.\n");
    exit(-1);
    }
  */
  qq = 1 / q;

  lam_max = 1 + qq + (2 * sqrt(qq));//LAMDA_MAX

  lam_min = 1 + qq - (2 * sqrt(qq));//LAMDA_MIN


  part = (lam_max - lam_min) / SEPARATE;

  lam[i] = lam_min;//INITIAL_VALUE

  while(i < SEPARATE + 1){

    r_lam[i] = q / (2 * M_PI) * (sqrt((lam_max - lam[i]) * (lam[i] - lam_min)
     ) / lam[i]);//RMT_DISTRIBUTION
    i++;
    lam[i] = lam[i - 1] + part;
  }

  // printf("ENTER INPUT_FILENAME");
  //scanf("%s",fname);
  sprintf(fname, "RMT_DISTRIBUTION.txt");
  fp = fopen(fname, "w");//INPUT_DATA

  i = 0;
  while(i < SEPARATE + 1){//INPUT_IN_RANGE
    // fprintf(fp,"%lf %lf\n",lam[i],r_lam[i]);
    fprintf(fp,"%lf %lf\n",lam[i],r_lam[i]);
    i++;
  }

```

```
  fclose(fp);

  return 0;
}
```

Listing B.2 CorrelationMatrix.c

```
/* 2 This program constructs Cross Correlation Matrix */
#include <stdio.h>
#include <string.h>
#include <stdlib.h>
#define NUM 1000 //MAX_COMPANIES
#define DATA 100 //NUMBER_COMPANIES
int main(void)
{
  FILE *rfp, *wfp, *rfp2, *rfp3,*wfp2;
  int label[NUM],a_label[DATA], dummy1[NUM];
  double dummy2[NUM];
  int i,j=0,count=0,temp;
  int flag[500];
  int size=0;
  int n;
  char file_code[200];
  for(i=0;i<500;i++) {
    flag[i]=0;
  }
  rfp=fopen("soukan2.txt", "r");
  if(rfp==NULL) {
    printf("read File open error");
    exit(1);
  }
  wfp=fopen("./R/com_label.txt", "w");
  if(wfp==NULL) {
    printf("write File open error\n");
    exit(1);
  }
  for(i=0;i<NUM;i++) {
    fscanf(rfp, "%d %d %lf\n",&label[i], &dummy1[i], &dummy2[i]);
    temp=label[i];
    //    printf("%d %d %lf\n",label[i],dummy1[i],dummy2[i]);

    if(flag[temp]==0) {
      flag[temp]++;
      count++;
      //fprintf(wfp,"%d\n", temp);
      if(count==DATA)
 break;
    }
   // printf("%d %d %lf\n",label[i],dummy1[i],dummy2[i]);
  }
  printf("100sya:%d\n",i);

  for(i=0;j<DATA;i++) {
    if(flag[i]==1) {
      a_label[j]=i;
      j++;
    }
  }
  for(j=0;j<DATA;j++){
    fprintf(wfp, "%d %d\n",j,a_label[j]);
  }
  fclose(wfp);
  fclose(rfp);
  /*-----------CREATE_LISTS--------------*/
  rfp2 = fopen("./file/correct3-1.txt", "r");
  if(rfp2==NULL) {
```

```
      printf("read2 File open error");
      exit(1);
    }
    rfp3 = fopen("./R/com_label.txt", "r");
    if(rfp3==NULL) {
      printf("read3 File open error");
      exit(1);
    }
    wfp2 = fopen("./R/highrank_list.txt", "w");
    if(wfp2==NULL) {
      printf("write2 File open error\n");
      exit(1);
    }
    j = 0;
    for(i = 0; i < DATA; i++){
      fscanf(rfp3, "%d %d", &dummy1[i], &label[i]);
      for(; j <= label[i]; j++){
        fscanf(rfp2, "%s %d", file_code, &temp);
      }
      fprintf(wfp2, "%s %d\n", file_code, temp);
    }
    fclose(wfp2);
    fclose(rfp2);
    fclose(rfp3);
    return 0;
}
```

Listing B.3 evv3_1.c

```
/* 3 This program Constructs Cross Correlation Matrix from Data and
 Computes Eigenvalues and Eigenvectors and Makes Histogram */
#include <stdio.h>
#include <stdlib.h>
#include <math.h>
#include <string.h>
#include "sslib5.h"

//#define  SIZE 483
//#define  TIME 1705              //TIME=L-1
#define  SEPARATE 50

#define LIST_FOLDER "file"  //Folder contains FileLists
#define LIST_FILE "correct3-1.txt"
#define DATA_FOLDER "tune"    //Folder that contains data in the FIleList
#define LOG_FOLDER "log"      //Folder taht contains log-difference

//./exe_fileSIZETIME
int main(int argc, char *argv[])
{
  int SIZE = atoi(argv[1]);
  int TIME = atoi(argv[2]);
  /*  //input SIZE,TIME by hands
   int SIZE, TIME;
   printf("N = ");
   scanf("%d", &SIZE);
   printf("L = ");
   scanf("%d", &TIME); */

  FILE *fp;
  FILE *fpw;
  FILE *dfp;                      //Input file for data(...txt)
  int i=0, j=0, k=0, k1;
  int  nr=1000000;                //Max_rotations(Jacobi algorithm)
  char filename[50], rfilename[50], lfilename[50];
  char vfilename[50];
  char logfilename[50];
```

```
  char dummyc, dummys[3][20], buff[256];

  //char tick_name[SIZE][40];     //Array for filename
  char **tick_name;
  double *r;             //Array for log-return histogram for Eigenvalues

  double eigenvalues[SIZE];  //Array for Eigenvalues

  double av[SIZE * SIZE];  //1-dimensional array for correlation matrix
  double **d;  //pointer to reserve memory to compute correlation matrix

  double readfile;                   //Varible to input filename
  double s = 0.0;                    //Mean
  double ss = 0.0;                   //Mean Square
  double sd = 0.0;                   //Standard Deviation
  double stock;                      //Stock Price
  double pre_stock;                  //Previous Price
  int time;

  // double x[SIZE][TIME];           //Array for input file
  double **x;

  double v[SIZE * SIZE];         //1-dimensional array for Eigenvalues
  double **eigenvectors;          //2-dimensional array for Eigenvector

  double eps=0.0000001;           //Conbersion criterion(Jacobi Algorithm)
  double *u;              //Inner Product for Eigenvector and Log-return

  int lcount;

  /* Reserve memory */
  /* --- Starting --- */

  d = malloc(sizeof(double *) * SIZE);
  if (d == NULL)
    exit(0);
  for (i = 0; i < SIZE; i++) {
    d[i] = malloc(sizeof(double) * SIZE);
  }

  eigenvectors = malloc(sizeof(double *) * SIZE);
  if (eigenvectors == NULL)
    exit(0);
  for (i = 0; i < SIZE; i++) {
    eigenvectors[i] = malloc(sizeof(double) * SIZE);
  }

  x = malloc(sizeof(double *) * SIZE);
  if (x == NULL)
    exit(0);
  for (i = 0; i < SIZE; i++) {
    x[i] = malloc(sizeof(double) * TIME);
  }

  tick_name = malloc(sizeof(double *) * SIZE);
  if (tick_name == NULL)
    exit(0);
  for (i = 0; i < SIZE; i++) {
    tick_name[i] = malloc(sizeof(double) * 40);
  }

  u = malloc(sizeof(double) * SIZE);
  r = malloc(sizeof(double) * (SIZE * TIME));
  /* --- END --- */

  /* INPUT DATA */
  printf("Start READING DATA\n");
```

```
/* --- STARTING --- */
sprintf(lfilename, "./%s/%s", LIST_FOLDER, LIST_FILE);
dfp = fopen(lfilename, "r");  // OPENING LISTFILE
if (dfp == NULL) {
  puts("File(list) open error.");
  exit(1);
}
for (i = 0; i < SIZE; i++) {
  readfile= fscanf(dfp, "%s%d", filename, &lcount)  // READ A LINE
  if (readfile == EOF) {    // IF NOT THE END OF LINE
    break;
  }

  //Write a fileneme on rfilename
  sprintf(rfilename, "./%s/%s", DATA_FOLDER, filename);

  fp = fopen(rfilename, "r"); // INPUT DATA FILE
  if (fp == NULL) {
    puts("File (read) open error.");
    exit(1);
  }
  strcpy(&tick_name[i][0], rfilename);

  sprintf(logfilename, "./%s/%s", LOG_FOLDER, filename);
  fpw = fopen(logfilename, "w");
  if (fp == NULL){
    puts("File (log) open error.");
    exit(1);
  }

  for (j = 0; j <= TIME; j++) {
    fscanf(fp, "%s %d %lf", dummys[0], &time, &stock); // READ DATA
    if(stock <= 0){
      printf("in the file named
  %s %d -th price has error\n", rfilename, j);
      exit(1);
    }

    fgets(buff, 256, fp);                    // discard after price
    if (j != 0) {                           //if not the first file
      x[i][j - 1] = log(stock / pre_stock);  //log-return from previous
  price
      fprintf(fpw,"%lf\n", x[i][j-1]);      //output log-return
      pre_stock = stock;
      s += x[i][j - 1];
      ss += x[i][j - 1] * x[i][j - 1];
    } else {
      pre_stock = stock;
    }
  }
  s /= TIME;                              //Calculate MEAN
  ss /= TIME;                             //Calculate MEAN Squares
  if (ss <= (s * s)) {
    printf("Error.\n");
    break;
  }
  sd = sqrt(ss - s * s);                   //Standard Deviation
  for (j = 0; j < TIME; j++) {
    x[i][j] = (x[i][j] - s) / sd;
  }
  s = 0;
  ss = 0;
  sd = 0;
  fclose(fp);
  fclose (fpw);
}
/* --- END --- */
```

```
  printf("END OF JOB\n\n");


  /*
   for(i = 0;i < SIZE;i++){
  printf("%s\n",&tick_name[i][0]);
  }
  */

  /*Histogram of log-return vectors */

  for(i = 0;i < SIZE;i++){
    for(j = 0;j < TIME ;j++){
      r[j + TIME * i] = x[i][j];
    }
  }
  printf("BEGIN WRITING Log-Return Vectors\n");
  sort(r,0,(SIZE*TIME)-1);
  sprintf(vfilename, "taisu_sort.txt");
  fpv = fopen(vfilename, "w");
  if(fpv == NULL){
    puts("File open error.");
    exit(1);
  }
  for(i = 0; i < SIZE; i++){
    for(j = 0;j < TIME;j++){
      fprintf(fpv,"%lf\n",r[j + TIME * i]);
    }
  }
  fclose(fp);
  histogram(r,SIZE*TIME,SEPARATE);

  /* --- END --- */
  printf("END OF JOB\n\n");

  /* Construct Cross Correlation matrix */
  printf("Start Construction Cross Correlation Matrix\n");
  /* --- STARTS --- */
  for (i = 0; i < SIZE; i++) {
    k1 = i * TIME;
    for (j = i; j < SIZE; j++) {
      for (k = 0; k < TIME; k++) {
        d[i][j] += x[i][k] * x[j][k];
      }
      d[i][j] /= TIME;
    }
    for (j = 0; j < i; j++) {
      d[i][j] = d[j][i];
    }
  }
  for (i = 0; i < SIZE; i++) {
    for (j = 0; j < SIZE; j++) {
      av[j + SIZE * i] = d[i][j];
    }
  }
  /* --- END --- */
  printf("End of Job n\n");

  //Write Eigenvalues and Eigenvectors named av
  printf("Starting Jacobi Algorithm\n");
  jacobi(av, v, SIZE, SIZE, &nr, eps);
  printf("End Jacobi\n\n");

  /* UTPUT EIGENVALUES */
  printf("START WRITING EIGENVALUES\n");
  /* --- STARTING --- */
  fpw = fopen("eigenvalues.txt", "w");  //MAKE FILE OF EIGENVALUES
```

```
    for(i = 0; i < SIZE; i++){
      for(j = 0; j < SIZE; j++){
        if(i == j){
          fprintf(fpw, "%d %s %lf\n",i,&tick_name[i][0], av[i*SIZE+j]);
          eigenvalues[i] =  av[i*SIZE+j];
        }
      }
    }
    fclose(fpw);
    /* --- END --- */
    printf("END \n\n");

    /* MAKE HISTOGRAM OF EIGENVALUES */
    //printf("START MAKING HISTOGRAM OF EIGENVALUES\n");
    //sort(eigenvalues,SIZE);
    //histogram(eigenvalues,SIZE,SEPARATE);
    //printf("END\n\n");

    /* OUTPUT EIGENVECTORS TO 2-DIMENSIONAL ARRAY */
    for(i = 0;i < SIZE;i++){
      for(j = 0;j < SIZE;j++){
        eigenvectors[i][j] = v[i * SIZE + j];
      }
    }
    printf("WRITING (%d)EIGENVALUES\n",SIZE);
    /* --- STARTING --- */
    for(i = 0;i < SIZE;i++){
      sprintf(vfilename, "./eigenvector/eigenvector%d.txt", i);
      fpv = fopen(vfilename, "w");
      if(fpv == NULL){
        puts("File open error.");
        exit(1);
      }
      for(j = 0;j < SIZE;j++){
        // eigenvectors[i][j] = v[i * SIZE +j];
// OUTPUT EIGENVECTORS TO FILE
        fprintf(fpv, "%d %s %lf\n",j,&tick_name[j][0],eigenvectors[j][i]);
      }
      fclose(fp);
    }
    /* --- END --- */
    printf("END OF JOB\n\n");

    /* RELEASE MEMORY */
    if (d){
      for (i = 0; i < SIZE; i++){
        if (d[i]){
          free(d[i]);
        }
      }
      free(d);
    }

    return 0;
}
```

Listing B.4 values_sort.c

```
/* 4 This program Sorts Eigenvectors */
#include <stdio.h>
#include <stdlib.h>
#include <math.h>
#include <string.h>

//#define SIZE 485

```

```
//./EXE_fileSIZE
int main(int argc, char *argv[])
{
  int SIZE = atoi(argv[1]);

  /*int SIZE;
    printf("SIZE = ");
    scanf("%d", &SIZE);*/

  int i = 0,j = 0,k = 0,l = 0,m = 0;//i,j,k:for loop
  FILE *dfp,*fpv;//file pointer

  double vector[SIZE][20];//3rd element of array to input eigenvectors
  char name[SIZE][40];    //2nd element of array to input file name
  int num[SIZE][5];       //1st element of array to input number

  char vfilename[40];
  double vhold;//sorting
  char namehold[40];//sorting

  printf("START\n");
  sprintf(vfilename, "eigenvalues.txt");
  dfp = fopen(vfilename,"r");
  if (dfp == NULL) {
    puts("eigenvalues.txt is not found.");
  }
  for(j = 0;j < SIZE;j++){//loop to read
    fscanf(dfp,"%d %s %lf",&num[j][0],&name[j][0],&vector[j][0]);
  }

    for(l = 0;l < SIZE - 1;l++){
      for(m = SIZE - 1;m > l;m--){
 if(vector[m - 1][0] < vector[m][0]){
   //===== Sorting eigenvectors =========
   vhold = vector[m][0];
   vector[m][0] = vector[m - 1][0];
   vector[m - 1][0] = vhold;
   //===== Sorting names =========
   strcpy(&namehold[0],&name[m][0]);
   strcpy(&name[m][0],&name[m - 1][0]);
   strcpy(&name[m - 1][0],&namehold[0]);
   //===== sorting numbers =========
   numhold = num[m][0];
   num[m][0] = num[m - 1][0];
   num[m - 1][0] = numhold;
 }
      }
    }

    sprintf(vfilename, "sort_values.txt");
    fpv = fopen(vfilename, "w");
    if(fpv == NULL){
      puts("FILE vectorsort is not found.");
      exit(1);
    }
    for(k = 0;k < SIZE;k++){
      fprintf(fpv, "%d %s %lf\n",num[k][0],&name[k][0],vector[k][0]);
    }
    fclose(fpv);
    printf("END\n\n");
  return 0;
}
```

Listing B.5 Sort_Eigenvector_Components.c

```
/* 5 This program Sorts eigenvector components */
#include <stdio.h>
#include <stdlib.h>
#include <math.h>
#include <string.h>

//#define SIZE 485

//./EXE_fileSIZE
int main(int argc, char *argv[])
{
  int SIZE = atoi(argv[1]);

  /*int SIZE;
    printf("SIZE = ");
    scanf("%d", &SIZE);*/

  int i = 0,j = 0,k = 0,l = 0,m = 0;//i,j,k:for loop
  FILE *dfp,*fpv;//file pointer

  double vector[SIZE][20];//3rd element of array to input eigenvectors
  char name[SIZE][40];     //2nd element of array to input file name
  int num[SIZE][5];        //1st element of array to input number

  char vfilename[40];
  double vhold;//sorting
  char namehold[40];//sorting

  printf("START\n");
  sprintf(vfilename, "eigenvalues.txt");
  dfp = fopen(vfilename,"r");
  if (dfp == NULL) {
    puts("eigenvalues.txt is not found.");
  }
  for(j = 0;j < SIZE;j++){//loop to read
    fscanf(dfp,"%d %s %lf",&num[j][0],&name[j][0],&vector[j][0]);
  }
      for(l = 0;l < SIZE - 1;l++){
      for(m = SIZE - 1;m > l;m--){
  if(vector[m - 1][0] < vector[m][0]){
    //===== Sorting eigenvectors =========
    vhold = vector[m][0];
    vector[m][0] = vector[m - 1][0];
    vector[m - 1][0] = vhold;
    //===== Sorting names =========
    strcpy(&namehold[0],&name[m][0]);
    strcpy(&name[m][0],&name[m - 1][0]);
    strcpy(&name[m - 1][0],&namehold[0]);
    //===== sorting numbers =========
    numhold = num[m][0];
    num[m][0] = num[m - 1][0];
    num[m - 1][0] = numhold;
  }
      }
    }
        sprintf(vfilename, "sort_values.txt");
    fpv = fopen(vfilename, "w");
    if(fpv == NULL){
      puts("FILE vectorsort is not found.");
      exit(1);
    }
    for(k = 0;k < SIZE;k++){
      fprintf(fpv, "%d %s %lf\n",num[k][0],&name[k][0],vector[k][0]);
    }
    fclose(fpv);
```

```
    printf("END\n\n");

    return 0;
}
```

Listing B.6 sslib2.h

```
/* include file for sslib */
#include <math.h>
#include <stdio.h>
#define FILTER 8 //filters for histograms

/* declare functions */
/*-----------------------------------------------------------*/
// print matrix
void mprint(char *s, double a[], int m, int n);

/*   EigenvalueEigenvectors by Jacobi Roration   */
int jacobi(double a[], double v[], int l, int m, int *nr, double eps);
/* Quick */
void sort(double x[],int left, int right);
/* Make histogram */
void histogram(double r[],int data,int separate);
/* Write files */
void  write_file(double min[],double max[],double per[],double part,int
     separate);

/* functions */
/*-----------------------------------------------------------*/
// print matrix
void mprint(char *s, double a[], int m, int n)
{
  int i, j, k;

  printf("matrix : %s\n", s);  // print matrix name
  for(i = k = 0; i < m; i++){
    for(j = 0; j < n; j++, k++){
      printf("%lf ", a[k]);
    }
    putchar('\n');
  }
}

/* EigenvalueEigenvectors by Jacobi Roration */
int jacobi(double a[], double v[], int l, int m, int *nr, double eps)
{
int n, i, j, r, c, k1, k2, s1, s2, s3, s5, s6;
double wmax, w, a1, a2, a3, t1, ta, si, co;  // mprint(" values0", a, 1, 1);
if(m < 2 || *nr < 1 || eps <= 0.0)
{
    //mprint(" values1", a, 1, 1);
    return 999;
  }
    n = 0;
    for(i = 0; i < m - 1; i++){
    k1 = i * l;
    for(j = i + 1; j < m; j++)
      if(a[k1 + j] != a[j * l + i])
{
        //printf("%d %d\n", i,j);
        //mprint(" values2", a, 1, 1);
        return 999;
      }
  }

  for(i = 0; i < m; i++){
```

```
    k1 = i * l;
    for(j = 0; j < m; j++)
      if(i != j)
        v[k1 + j] = 0.0;
    v[k1 + i] = 1.0;
  }
  while(1)
{
    wmax = 0.0;
    for(i = 0; i < m; i++){
      k1 = i * l;
      for(j = i + 1; j < m; j++){
        w = fabs(a[k1 + j]);
        if(w > wmax){
          wmax = w;
          r = i;
          c = j;
        }
      }
    }

    /*
     printf("loop_n : n = %d\n",n);
     printf("Converg_bound: eps = %.9f\n",eps);
     printf("abs_max: wmax = %.9f\n",wmax);
     printf("i = %d,j = %d \n",r ,c);
     */
    if(wmax <= eps){
      *nr = n;
      return 0;
    }

    if(n >= *nr){
      *nr = n;
      return 1;
    }

    k1 = r * l;
    k2 = c * l;
    s1 = k1 + r;
    s2 = k2 + c;
    s3 = k1 + c;
    n++;
    a1 = a[s1];
    a2 = a[s2];
    a3 = a[s3];
    t1 = fabs(a1 - a2);
    ta = 2.0 * a3 / (t1 + sqrt(t1 * t1 + 4.0 * a3 * a3));
    if(a1 < a2) ta = - ta;
    co = sqrt(1.0 / (ta * ta + 1.0));
    si = ta * co;

    for(i = 0; i < m; i++){
      s5 = i * l + r;
      s6 = i * l + c;
      w = v[s5];
      v[s5] = w * co + v[s6] * si;
      v[s6] = - w * si + v[s6] * co;
      if(i != r && i != c){
        w = a[s5];
        a[s5] = w * co + a[s6] *si;
        a[s6] = - w * si + a[s6] * co;
        a[k1 + i] = a[s5];
        a[k2 + i] = a[s6];
      }
    }

```

```
      a[s1] = a1 * co * co + a2 * si * si + 2.0 * a3 * co * si;
      a[s2] = a1 + a2 - a[s1];
      a[s3] = 0.0;
      a[k2 + r] = 0.0;

    }
  }

//Quick sort
void sort(double x[ ], int left, int right)
{
    int i, j;
    double pivot;
    double temp;
  reference point
    i = left;                        /* suffix to min_element of array to sort
     */
    j = right;                       /* suffix to MAX_element of array to sort
     */

    pivot = x[(left + right) / 2]; /* Choose central value as reference pt */

    while (1) {                      /* infinite loop */

        while (x[i] < pivot)         /* until value larger than pivot appears
     */
            i++;                     /* increase i */

        while (pivot < x[j])         /* until value smaller than pivot appears
     */
            j--;                     /* decrease j */
        if (i >= j) {                 /* if i >= j */
            break;                   /* exit infinite loop */
    }
        temp = x[i];                 /* exchange x[i] and x[j] */
        x[i] = x[j];
        x[j] = temp;

        i++;                         /* next data */
        j--;
    }
    if (left < i - 1)                /* if 2 or more elements exist in lhs of
     reference */
        sort(x, left, i - 1);       /* Q sort the left array */
    if (j + 1 <  right)              /* if 2 or more elements exist in rhs of
     reference */
        sort(x, j + 1, right);      /* Q sort the right array */
}


/* make histogram */
void histogram(double r[],int data,int separate)
{

  double part,up = 0.0,per[separate + 1];
  int i = 0,j = 0,c = 0,ck=0;
  double min[separate],max[separate],sum=0;
  double smax = -10000,smin = 100000,ss = 0.0;
  int box[separate + 2];
  int count;
  int ck_sum=0;

  double d = 0.0,sd = 0.0,ave = 0.0;//variance standard deviation mean;
  a
  int filter_count =0;

  for(i = 0;i < data;i++){// find max and min
```

```
    if(smax < r[i]){
      if (r[i] < FILTER){      //upper level filter
      }else{
        //printf("%lf\n",r[i]);
        filter_count++;
      }
    }
    if(smin > r[i]){
      if(r[i] > -FILTER){      //lower level filter
        smin = r[i];
      }else{
        //printf("%lf\n",r[i]);
        filter_count++;
      }
    }
    ave += r[i];
    ss += r[i] * r[i];
  }

  smax = FILTER;
  smin = -FILTER;

  part = (smax - smin) / separate;
  min[j] = smin;
  max[j] = min[j] + part;

  ck = 0;
  /*
   for(i = 0;i < count;i++){
   if(r[i] < smin){//lower side
   ck++;
   }
   }
   */
  box[c] = ck;
  c++;

  ck = 0;

  for(i = 0;i < separate+1;i++){//in the of the region
    if(i == 0){                     //fix min
      min[i] = smin;
    }
    else{
      min[i] = max[i - 1];
    }
    max[i] = min[i] + part;
    up = max[i];//for outside ewgion
    ck = 0;//reset accordng to new box

    for(j = 0;j < data;j++){
      if(-FILTER <= r[j] && r[j] <= FILTER)
        if(min[i] <= r[j] && r[j] < max[i]){
          //  if(ck < data/10){
          ck++;
          /* }else{
           filter_count++;
           } */
        }
    }

    box[c] = ck;
    c++;
  }

  ck = 0;
  /*
```

```
    for(i = 0;i < count;i++){
    if(up <= r[i]){//outside region upper side
    ck++;
    }
    }
    */
  box[c] = ck;

  printf("activate filter %d-count \n",filter_count);

  count = data - filter_count;

  ave /= count;//mean
  ss /= count;
  d = ss - ave * ave;//variance
  sd = sqrt(d);//standard deviation

  printf("mean:%lf   variance:%lf   stansardDev:%lf\n",ave,d,sd);
  printf("max:%lf   min:%lf\n",smax,smin);
  printf("n_data:%d   n_parts:%d  width_part:%lf\n",data,separate,part);


  for(i = 0;i < separate + 2;i++){
    per[i] = (double)box[i] / (double)count / part;// ProbabilityDensityDistribution
    sum += per[i] * part; //add_area
  }
  //-----------------------------------------------

  for(i = 0;i < separate + 2;i++){
    ck_sum += box[i];
  }

  printftotal area%lf \n",sum);
  printf("n_count%d \n",ck_sum);

  write_file(min,max,per,part,separate);

}



void  write_file(double min[],double max[],double per[],double part,int separate)
{

  FILE *fp;
  int i=0;
  double ave[separate],dif;
  char fname[30] = "taisu_vector.txt";
  dif = part / 2;

  //printf("Input White File Name:");
  //scanf("%s",fname);
  //fname = "taisu_vector.txt";
  fp = fopen(fname, "w");
  if(fp == NULL){
    printf("File open error.\n");
  }

  // order to write is min mid prob
  //            max mid prob
  fprintf(fp,"%lf %lf %lf\n",min[i] - 2*dif,(2*min[i]-3*dif)/2,0.0);//fix min graph
  fprintf(fp,"%lf %lf %lf\n",min[i] - dif,(2*min[i]-3*dif)/2,0.0);

```

```
    fprintf(fp,"
      %lf %lf %lf box[%d]\n",min[i] - dif,(2*min[i]-dif)/2,per[i],i);//smaller
      than the region
    fprintf(fp,"%lf %lf %lf box[%d]\n",min[i],(2*min[i]-dif)/2,per[i],i);

    while(i < separate){//within the region
      ave[i] = min[i] + dif;
      fprintf(fp,"%lf %lf %lf box[%d]\n",min[i],ave[i],per[i+1],i+1);
      fprintf(fp,"%lf %lf %lf box[%d]\n",max[i],ave[i],per[i+1],i+1);
      i++;
    }

    i++;

    fprintf(fp,"
      %lf %lf %lf box[%d]\n",max[i-2],(2*max[i-2]+dif)/2,per[i],i);;//larger
      than the region
    fprintf(fp,"
      %lf %lf %lf box[%d]\n",max[i-2]+dif,(2*max[i-2]+dif)/2,per[i],i);
    fprintf(fp,"
      %lf %lf %lf\n",max[i-2]+dif,(2*max[i-2]+3*dif)/2,0.0);;//fix max graph
    fprintf(fp,"%lf %lf %lf\n",max[i-2]+2*dif,(2*max[i-2]+3*dif)/2,0.0);
    fclose(fp);
}
```

Appendix C
Program for the RMT-test

The outline of the job executed by the program shown below is as follows:

(1) To cut the input data into many terms of equal length, $N \times L$ so that $L = Q \times N$, beginning from the first item (e.g., Term 1 contains January 4–13, 2021 data for TOPIXcore30, for $N = 200$).
(2) To construct a correlation matrix, $C = GG^t$, which is an N-by-N matrix, made by G (an N-by-L matrix) and its transverse matrix G^t (an L-by-N matrix), both made by cutting the piece (of length $N \times L$) into sub-pieces of length L.
(3) To compute the eigenvalues of C by using Jacobi's diagonalization algorithm and output them as files named "eigenvalue0001.dat," "eigenvalue0002.dat,"..., "eigenvalue0036.dat," if there are 36 terms in the input file.
(4) To calculate the histogram of eigenvalue distribution and output them as files named "histogram0001.dat," "histogram0002.dat,"..., "histogram0036.dat," if there are 36 terms.
(5) To compute the theoretical distribution curve and output them as a file named "PRMT.dat."
(6) To compute the randomness as k-th moments for $k = 2, 3, 4, 5, 6$ and output them as files named "RMTrandomness_0001_term.dat," "RMTrandomness_0002_term.dat,"..., "RMTrandomness_0036_term.dat," if there are 36 terms.

****
****

⟨⟨Description of the main program called "Sample.c"⟩⟩

The following is our program, named "Samle.c," used to analyze a price time series of length $N \times N \times Q$. This program analyzes an input data to do the following jobs.

M. Tanaka-Yamawaki, Y. Ikura, *Principal Component Analysis and Randomness Test for Big Data Analysis*, Evolutionary Economics and Social Complexity Science 25, https://doi.org/10.1007/978-981-19-3967-9

"Input.txt" is a name of data file that contains the input data of a time series of an index price (such as TOPIX) per 1 second for each year. Each line has 2 items (such as time price). For example, the 2021 data begin as follows: ⟨contents of the input file⟩

```
20140106090005693.45
20140106090006692.95
20140106090007693.25
20140106090008693.03
20140106090009693.32
...
```

Listing C.1 Sample.c

```c
#include<stdio.h>
#include<stdlib.h>
#include<string.h>
#include<math.h>

#define INPUT "input.txt"           /* input file */
#define Q 3                         /* Q */
#define N 300                       /* N */
#define L (Q * N)                   /* L */
#define EPSILON 10e-30              /* minimum number to be regarded as zero
    */
#define THR 2000000                 /* maximum number of iterations */
#define DN  20                      /* division to draw a histogram */

void initial_processing(
      int *mmax                     /* number of terms */
      );
void prmt_curve();
void make_matrix(
      int m,                        /* current term */
      double x[][N+1],              /* N-by-N matrix */
      char str[][256]               /* starting-ending date of the current
      term */
          );
void eigenvalue(
     int m,                         /* current term */
     double x[][N+1],               /* N*N matrix */
     double maxmin[],               /* max and min eigenvalues */
     char str[][256],               /* starting-ending date of the current
      term */
     FILE *fp0                      /* pointer to output delta_lambda.dat */
     );
void histogram(
          int m,                    /* current term */
          double x[][N+1],          /* N*N matrix */
          double maxmin[]           /* max and min eigenvalues */
          );
void randomness(
     int m,                         /* current term */
```

```
    double x[][N+1]                   /* N*N matrix */
    );
void find_max(
        double *max,                  /* maximum element in matrix x */
        int *k,                       /* row where the maximum element resides */
        int *l,                       /* column where the maximum element resides */
        double x[][N+1]);             /* N*N matrix */

int main()
{
  int m;                              /* variable for term */
  int mmax;                           /* number of terms */
  double maxmin[2];                   /* storage of max and min eigenvalues */
  double x[N+1][N+1];                 /* N*N matrix */
  char str[3][256];                   /* storage for the first row of Input.txt */
  FILE *fp;
  if ((fp = fopen("delta_lambda.dat", "w")) == NULL){
    printf(File open error\n);
    return -1;
  }
  initial_processing(&mmax);
  prmt_curve();
  for (m = 1; m <= mmax; m++)  {
    printf("term %d / %d\n", m, mmax);
    make_matrix(m, x, str);
    eigenvalue(m, x, maxmin, str, fp);
    histogram(m, x, maxmin);
    randomness(m, x);
  }
  fclose(fp);
  return 0;
}

void initial_processing(int *mmax)
/*
This sub-routine computes the integer part of the ratio of the number of data in the input file
Input.txt and the size of each block N*N*Q, and outputs the result as the number of blocks
to the variable mmax.
 */
{
  int dummy;
  char str[256];
  FILE *fp;
  sprintf(str, "wc -l %s", INPUT);
  if ((fp = popen(str, "r")) == NULL )
  {
    printf("File open Error\n");
    exit(1);
  }
  fscanf(fp, "%d\n", &dummy);
  pclose(fp);
  *mmax = (dummy - 1) / L / N ;
}

void prmt_curve()
/*
This subroutine outputs the data to draw the theoretical distribution function in Eq.(IV-14)
to the output file named PRMT.dat
 */
{
  int i;
```

```
  double u, du, dl, tmp, length, lambda_p, lambda_m;
  FILE *fp;
  lambda_p = 1.0 + 1.0/Q + 2.0 * sqrt(1.0/Q);
  lambda_m = 1.0 + 1.0/Q - 2.0 * sqrt(1.0/Q);
  length   = lambda_p - lambda_m;
  dl       = length / DN;
  if ((fp = fopen("PRMT.dat", "w")) == NULL)
  {
    exit(1);
  }
  du = length / 5000.0;
  for (i = 0; i <= 5000; i++)
  {
    u   = lambda_m + i*du;
    tmp = sqrt( ( lambda_p - u ) * ( u - lambda_m ) ) / u * 0.5 * Q / M_PI;
    fprintf(fp, "%g %g\n", u, tmp);
  }
  fclose(fp);
}

void make_matrix(int m, double x[][N+1],char str[3][256])
/*
This subroutine outputs the rectangular matrix G in Eq.(IV-4), from which the
       correlation matrix
C=GG^t is constructed
*/
{
  int i, j, k;
  double tmp[2], sum[2];
  char dummy[256];
  static double  z[N+1][L+1];
  FILE *fp;
  if ((fp = fopen(INPUT, "r")) == NULL)
  {
    printf("FILE OPEN ERROR\n");
    exit(1);
  }
  for (i = 1 ; i <= N*L*(m-1); i++)
  {
      fscanf(fp, "%s %lf\n", str[1], &tmp[0]);
  }
  fscanf(fp, "%s %lf\n", str[1], &tmp[0]);
  for (i = 1; i <= N; i++)
  {
    for (j = 1; j <= L; j++)
    {
      fscanf(fp, "%s %lf\n", dummy, &tmp[1]);
      if (i== N && j==L-1){
  strcpy(str[2], dummy);
      }
      z[i][j] = log(tmp[1]) - log(tmp[0]);
      tmp[0] = tmp[1];
    }
  }
  fclose(fp);
  for (i = 1; i <= N; i++)
  {
    sum[0] = sum[1] = 0.0;
    for (k = 1; k <= L; k++)
    {
      sum[1] = sum[1] + z[i][k];
    }
    sum[1] = sum[1] / L;

    for (k = 1; k <= L; k++)
    {
      z[i][k] = z[i][k] - sum[1];
```

```
        sum[0]  = sum[0] + z[i][k]*z[i][k];
      }
      sum[0] = 1.0 / sqrt(sum[0]);
      for (k = 1; k <= L; k++)
      {
        z[i][k] = z[i][k] * sum[0];
      }
    }
    for (i = 1; i <= N; i++)
    {
      for (j = i; j <=N; j++)
      {
        sum[0] = 0.0;
        for (k = 1; k <= L; k++)
        {
    sum[0] = sum[0] + z[i][k] * z[j][k];
        }
        x[i][j] = x[j][i] = sum[0];
      }
    }
}

void eigenvalue(int m, double x[][N+1], double maxmin[], char str[][256], FILE *fp0)
/*
This subroutine diagonalizes given correlation matrix x[N+1] [N+1] by means of Jacobi algorithm
and returns the result to the diagonal elements of the same matrix x[N+1] [N+1] then outputs the
same result to a file named eigenvalueXXXX.dat where m shows the block number.
The output file named delta_lambda.dat contains (starting-ending date, ) for all terms.
 */
{
  int n, k, l, i, j;
  double max, theta, lambda_p;
  double y[N+1][N+1];
  FILE *fp1;

  max = 1.0;
  n   = 0;
  lambda_p = 1.0 + 1.0/Q + 2.0 * sqrt(1.0/Q);
  while (max > EPSILON)
  {
    n++;
    if (n > THR)
    {
      break;
    }
    find_max(&max, &k, &l, x);
    if (fabs(x[k][k] - x[l][l]) >= EPSILON)
    {
      theta = 2.0 * x[k][l] / (x[k][k] - x[l][l]);
      theta = 0.5 * atan(theta);
    }
    else
    {
      theta = 0.25*M_PI;
    }
    for (i = 1; i <= N; i++)
    {
      for (j = 1; j <= N; j++)
      {
  y[i][j] = x[i][j];
      }
    }
```

```
    for (j = 1; j <= N; j++)
    {
      y[k][j] =  x[k][j] * cos(theta) + x[l][j] * sin(theta);
      y[j][k] =  y[k][j];
      y[l][j] = -x[k][j] * sin(theta) + x[l][j] * cos(theta);
      y[j][l] =  y[l][j];
    }
    y[k][k] = x[k][k] * cos(theta)*cos(theta)
      + x[l][l] * sin(theta) * sin(theta) + x[k][l] * sin(2.0 * theta);
    y[l][l] = x[k][k] * sin(theta)*sin(theta)
      + x[l][l] * cos(theta) * cos(theta) - x[k][l] * sin(2.0 * theta);
    y[k][l] = y[l][k] = 0.0;
    for (i = 1; i <= N; i++)
    {
      for (j = 1; j <= N; j++)
      {
  x[i][j] = y[i][j];
      }
    }
  }
  sprintf(str[0], "eigenvalue%04d.dat", m);
  if ((fp1 = fopen(str[0], "w")) == NULL)
  {
    exit(1);
  }
  maxmin[0] = 0;
  maxmin[1] = 10000;
  for (i = 1; i <= N; i++)
  {
    fprintf(fp1, "%g\n", x[i][i]);
    if (maxmin[0] < x[i][i])
    {
      maxmin[0] = x[i][i];
    }
    if (maxmin[1] > x[i][i])
    {
      maxmin[1] = x[i][i];
    }
  }
  fclose(fp1);

  fprintf(fp0, "%s-%s %lf\n", str[1], str[2], maxmin[0] - lambda_p);
}

void find_max(double *max, int *k, int *l, double x[][N+1])
/*
This sub-routine finds the maximum element of matrix x and outputs its (value, row, column) to
the three variables (max, k, l), respectively
*/
{
  int i, j;
  double dummy, r;
  r = -1.0;
  for (i = 1; i < N; i++)
  {
    for (j = i+1; j <= N; j++)
    {
      dummy = fabs(x[i][j]);
      if ( r < dummy )
      {
  *max = r = dummy ;
  *k=i;
  *l=j;
      }
    }
  }
```

```
}

void histogram(int m, double x[][N+1], double maxmin[])
/*
This subroutine computes data to draw histogram and output them to a file named
histogramXXXX.dat where XXXX shows the block number m
 */
{
  int i, j, tmp, number[DN+1];
  double lambda_p, lambda_m, dl, position;
  char str[256];
  FILE *fp;
  lambda_p = maxmin[0] + EPSILON;
  lambda_m = maxmin[1] - EPSILON;
  dl = (lambda_p - lambda_m) / DN;
  for (j = 1; j <= DN; j++)
  {
    number[j] = 0;
  }
  for (i = 1; i <= N; i++){
    for (j = 1; j <= DN; j++){
      if (j!=DN)
      {
  if (x[i][i] >= lambda_m + (j-1)*dl && x[i][i] < lambda_m + j*dl)
  {
    number[j] ++ ;
  }
      }
      else
      {
  if (x[i][i] >= lambda_m + (j-1)*dl)
        {
    number[j] ++ ;
  }
      }
    }
  }
  sprintf(str, "histogram%04d.dat", m);
  if ((fp = fopen(str, "w")) == NULL)
  {
    exit(1);
  }
  tmp = 0;
  for (j = 1; j <= DN; j++)
  {
    tmp = tmp + number[j];
  }
  fprintf(fp,"%15.8e %15.8e\n",lambda_m,0.0);
  for (j = 1; j <= DN; j++)
  {
    position = lambda_m + (j - 1) * dl;
    fprintf( fp, "%15.8e %15.8e\n", position, number[j]/dl/tmp);
    position = lambda_m + j * dl;
    fprintf( fp, "%15.8e %15.8e\n", position, number[j]/dl/tmp);
  }
  fprintf( fp, "%15.8e %15.8e\n", lambda_p, 0.0);
  fclose(fp);
}

void randomness(int m, double x[][N+1])
/*
This subroutine computes the k-th deviation Dk from the random walk and outputs the result to
a file named DRMT_XXXX_term.dat the content of which is the inverse of the randomness.
*/
```

```
{
  int i, k;
  double tmp, mu[7], muR[7];
  char str[256];
  FILE *fp;
  for (k = 1; k <= 6; k++)
  {
    mu[k] = 0.0;
    for (i = 1; i <= N; i++)
    {
      mu[k] += pow(x[i][i], k);
    }
    mu[k] = mu[k] / N;
  }
  muR[1] = 1.0;
  muR[2] = 1.0 + 1.0 / Q;
  muR[3] = 1.0 + 3.0 / Q + 1.0 / (Q*Q);
  muR[4] = 1.0 + 6.0 / Q + 6.0 / (Q*Q) + 1.0 / (Q*Q*Q);
  muR[5] = 1.0 + 10. / Q + 20. / (Q*Q) + 10. / (Q*Q*Q) + 1.0 / (Q*Q*Q*Q);
  muR[6] = 1.0 + 15. / Q + 50. / (Q*Q) + 50. / (Q*Q*Q) + 10. / (Q*Q*Q*Q) +
      1.0 /(Q*Q*Q*Q*Q);
  sprintf(str,"DRMT_%04d_term.dat",m);
  if ((fp = fopen(str, "w")) == NULL)
  {
    exit(1);
  }
  for (k = 1; k <= 6; k++)
  {
    tmp = muR[k] / (mu[k] - muR[k]);
    fprintf(fp, "%d %15.8e\n", k, 1.0/tmp );
  }
  fclose(fp);
}
```

To compile and run this program, type the following:

```
% gcc Sample.c
% time ./a.out
```

The following message will appear, showing success. Simultaneously, output files are created.

```
term 1 / 2
term 2 / 2
./a.out 350.80s user 4.26s system 95% cpu 6:10.27 total
```

Let us check the created files by using "ls."

```
% ls
```

Files named "*.dat" are created.

```
Sample.c sub.c Input.txt a.out PRMT.dat delta_lambda.dat eigen-
value0001.dat eigenvalue0002.dat histogram0001.dat histogram0002.dat
DRMT_0001_term.dat DRMT_0002_term.dat
```

The file named "PRMT.dat" contains data to draw the theoretical curve of Eq. (4.14) in the text. The file named "delta_lambda.dat" contains starting–ending date of Term m and the value of the $\Delta\lambda$ for all the terms. The 3 files named "eigenvaluesXXXX.dat," "histogramXXXX.dat," and "DRMT_XXXX_term.dat" are the output files created for each term.

Let us check the content of "delta_lambda.dat."

```
% cat delta_lambda.dat
000001-270000 0.445578 270001-540000 0.463399
```

Each line shows (starting–ending date, $\Delta\lambda$) for each term. The file "eigenvalueXXXX.dat" contains all the eigenvalues in Term XXXX.

```
% tail eigenvalue0001.dat
0.750749
0.301738
0.232319
2.31665
1.84666
1.22346
1.20985
1.41795
1.41007
0.28578
```

The eigenvalue distributions of the Term XXXX are stored in "histogramXXXX.dat" which can be viewed, together with the theoretical curve stored in "PRMT.dat" by using gnuplot as follows:

```
gnuplot> set xlabel 'eigenvalue' ; set ylabel 'probability density'
gnuplot> set multiplot layout 1,2
multiplot> plot 'histogram0001.dat' with line, 'PRMT.dat' with line
multiplot> plot 'histogram0002.dat' with line, 'PRMT.dat' with line
```

Then, the output can be obtained as shown in Fig. C.1.

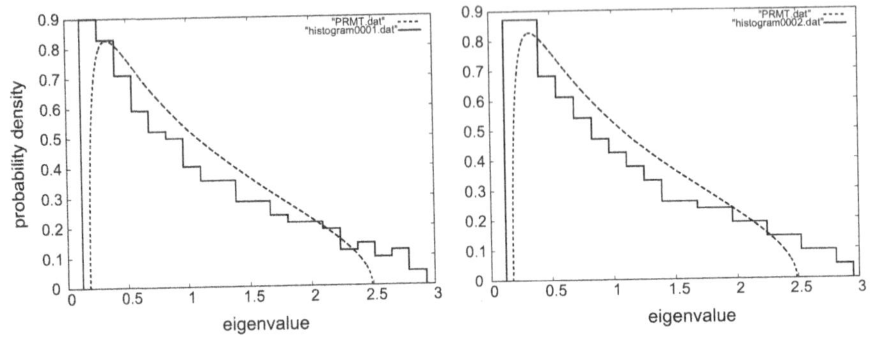

Fig. C.1 Histogram visualization with gnuplot

The output file named "DRMT_XXXX_term.dat" contains (k, D_k) for $k = 1-6$, where D_k defined in Eq. (4.19) represents the inverse of the randomness.

```
% cat DRMT_0001_term.dat
1 -1.11022302e-15
2 1.20788327e-01
3 3.05315669e-01
4 5.40694316e-01
5 8.29443486e-01
6 1.18949630e+00
```

The output of this program for TOPIX index in 2011.1–2012.5 is shown in Appendix D, and the output for TOPIXcore30 index in 2014 is shown in Appendix E

A sub-program named sub.c is prepared to output an artificial random data, for a reader who wishes to try the "Sample.c" without preparing real data for the input file "Input.txt."

By running the sub.c, the output file named "Input.txt" is created that contains two numbers each column: (the number of iterations rand()) to be used in the main program.

Listing C.2 sub.c

```
#include <stdio.h>
#include <stdlib.h>
#include <time.h>

#define N 300
#define Q 3
#define L Q*N
#define MAX_TERM 2

int main( void )
```

```
{
  int i,imax;
  double num;

  imax = N * L * MAX_TERM;
  srand( time( NULL ) );

  for (i = 1; i <= imax+1; i++){
    num = (double) rand() / RAND_MAX;
    printf("%d\t%f\n", i, num);
  }

  return 0;
}
```

Let us compile sub.c by the following command to create a.out:

```
% gcc sub.c
```

Run a.out and output the result into "Input.txt."

```
% ./a.out > Input.txt
```

Appendix D
RMT-test Applied on TOPIX Index Time Series in 2011.1–2012.5

The output of the program sample.c in Appendix C is presented here, for the case of using the second-by-second data of TOPIX index in the period of January 12, 2011 to June 11, 2012. The data is split into 48 pieces of length 120,000, corresponding to the parameters $N = 200$ and $Q = 3$ so that the size of rectangular matrix G is 200×600. There are 12 figures, Figs. D.1, D.2, D.3, D.4, D.5, D.6, D.7, D.8, D.9, D.10, D.11, and D.12, each of which contains the results of four terms. For example, Fig. D.1 contains the results from Term 1 to Term 4, and so on. The days included in each Term is a little less than 2 weeks, excluding long vacations. Since one working day contains 5 hours: 9:00-11:30, 11:30-15:00, there are 180,000 seconds per day. Therefore, at least 7 working days are necessary to make one term. For instance, Term 1 is made of January 12–24, 2011.

In Fig. D.1, there are two figures of Term 1. The left-hand figure shows the eigenvalue distribution with two outputs, λ_{max}, the largest eigenvalue of the correlation matrix and D_6, the randomness measure defined in Eq. (4.33). The right-hand figure shows the daily prices of Term 1 and Term 2. The entire period shown in Fig. D.1 lies in the random walk mode. This can be recognized easily by observing the histogram of the eigenvalue distribution in solid line, which almost matches the RMT-theoretical distribution of Eq. (2.28) in dotted line. The same is also true for the other terms in Fig. D.1. The randomness measure D_6, is very small throughout this period. However, the situation suddenly changes after Term 6 (March 7-17, 2011) in Fig. D.2 when a dramatic change occurs in the daily price. At the same time, the largest eigenvalue λ_{max} jumps to 7.16 and $D_6 = 52.51$. Term 6 includes the March 11 Great East Japan Earthquake occurred. Note that the deviation from the random walk of the second-by-second price fluctuation and the drop of the daily price occurred simultaneously without any prior events. This is due to external factors. Such changes cannot be predicted and this type of price fall can be called as exogenous type of price declines.

On the other hand, there are different types of price decline associated with a mild fall of randomness in the second-by-second price fluctuation. In the period in

M. Tanaka-Yamawaki, Y. Ikura, *Principal Component Analysis and Randomness Test for Big Data Analysis*, Evolutionary Economics and Social Complexity Science 25, https://doi.org/10.1007/978-981-19-3967-9

this Appendix D, we observe two such examples. The first example can be seen in Term 28 (Nov.4–11, 2011) in Fig. D.7 and the second example in Term 42 (Mar. 13–23, 2012) in Fig. D.11. Both cases accompany the market decline in the following Terms. The remaining figures (Figs. D.8, D.9, D.10, and D.12, as well as Figs. D.3, D.4, D.5, and D.6) can be regarded as the mode of the random walk.

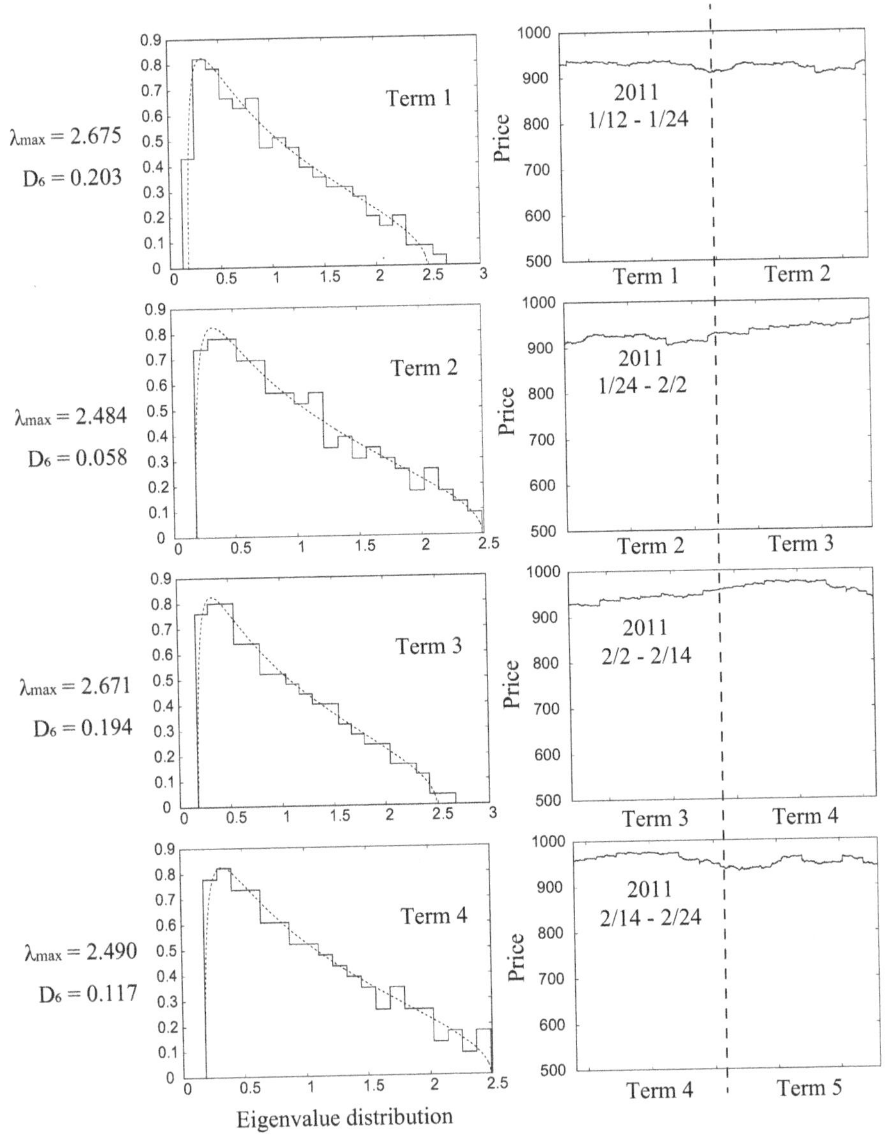

Fig. D.1 Output results for Term 1–4 are shown; λ_{max}, D_6, Eigenvalue Distribution, and Daily index

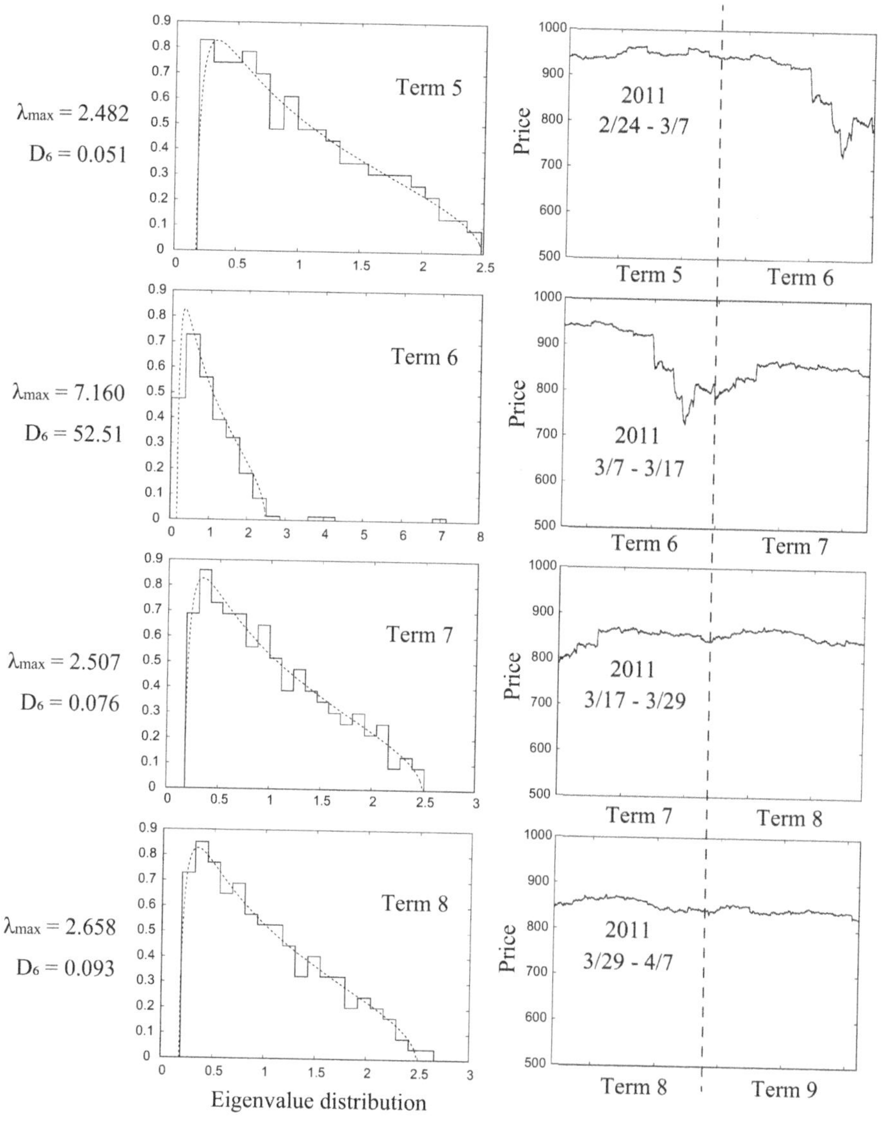

Fig. D.2 Output results for Term 5–8 are shown; λ_{max}, D_6, Eigenvalue Distribution, and Daily index

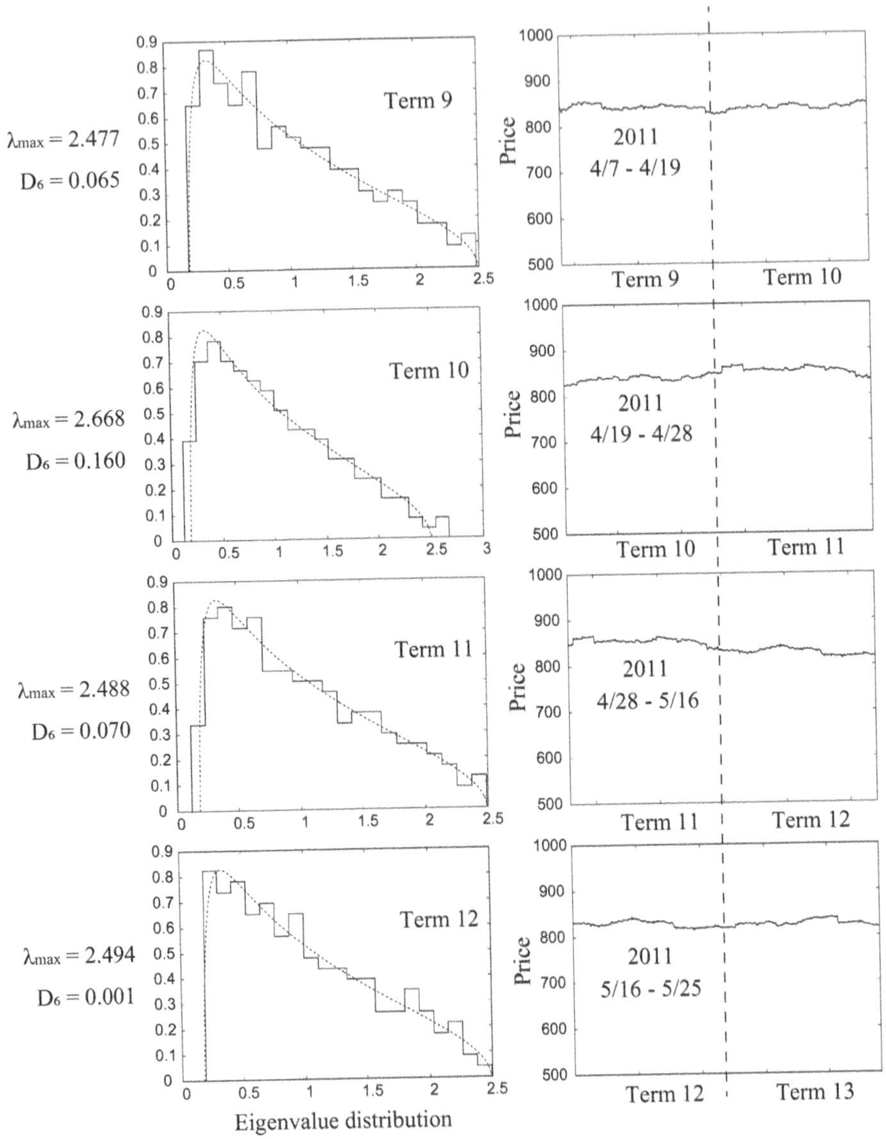

Fig. D.3 Output results for Term 9–12 are shown; λ_{max}, D_6, Eigenvalue Distribution, and Daily index

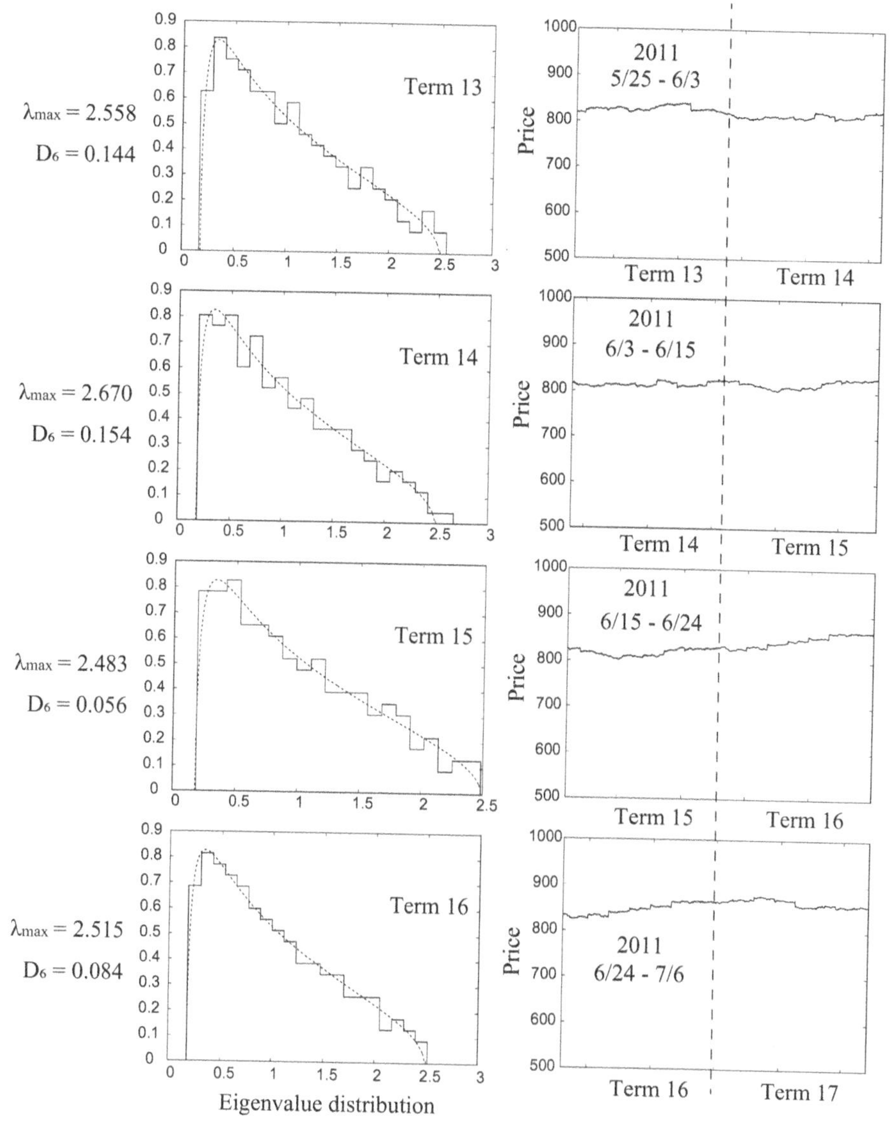

Fig. D.4 Output results for Term 13–16 are shown; λ_{max}, D_6, Eigenvalue Distribution, and Daily index

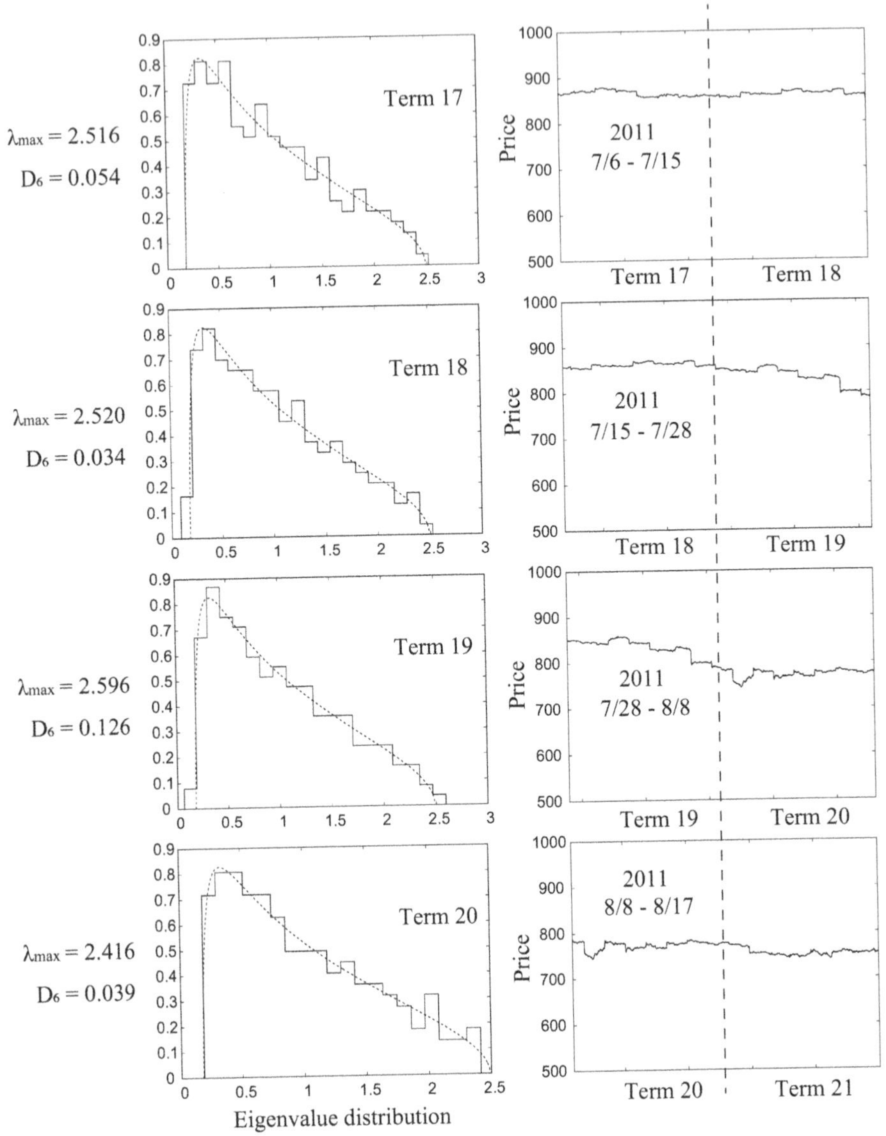

Fig. D.5 Output results for Term 17–20 are shown; λ_{max}, D_6, Eigenvalue Distribution, and Daily index

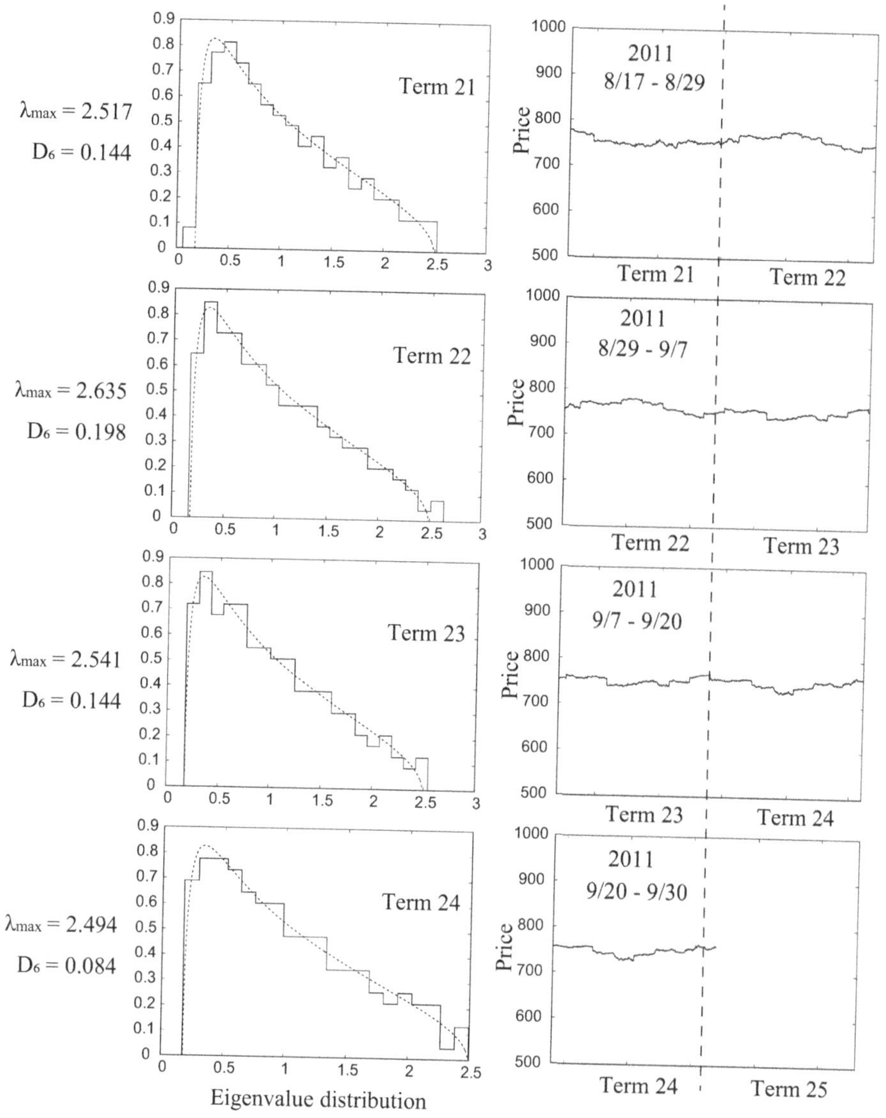

Fig. D.6 Output results for Term 21–24 are shown; λ_{max}, D_6, Eigenvalue Distribution, and Daily index

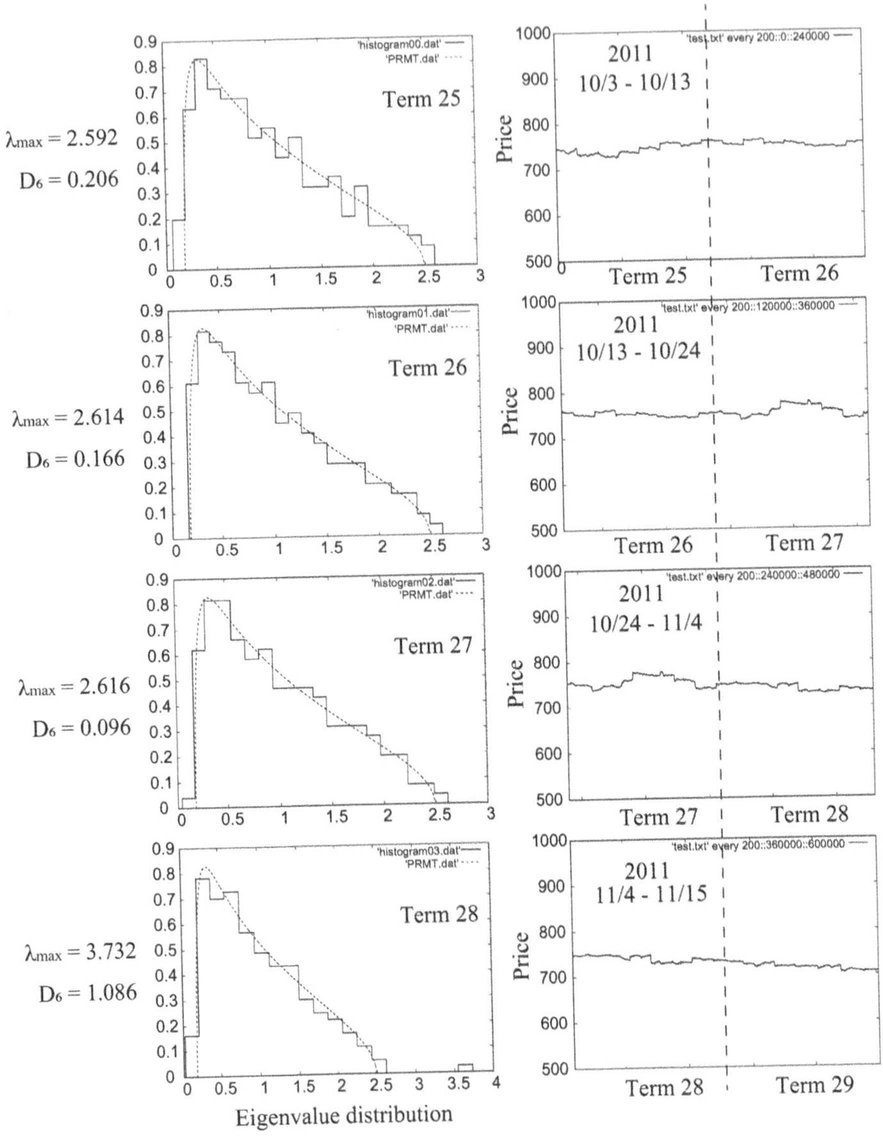

Fig. D.7 Output results for Term 25–28 are shown; λ_{max}, D_6, Eigenvalue Distribution, and Daily index

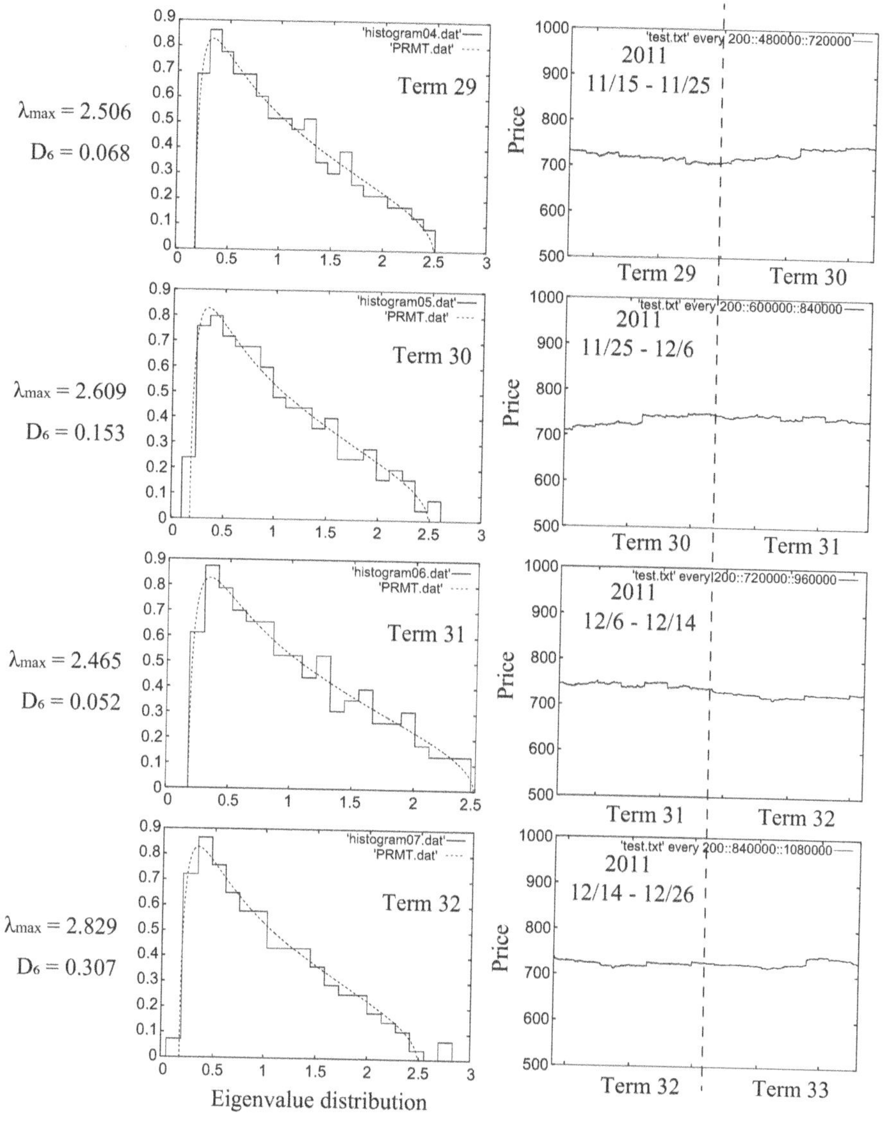

Fig. D.8 Output results for Term 29–32 are shown; λ_{max}, D_6, Eigenvalue Distribution, and Daily index

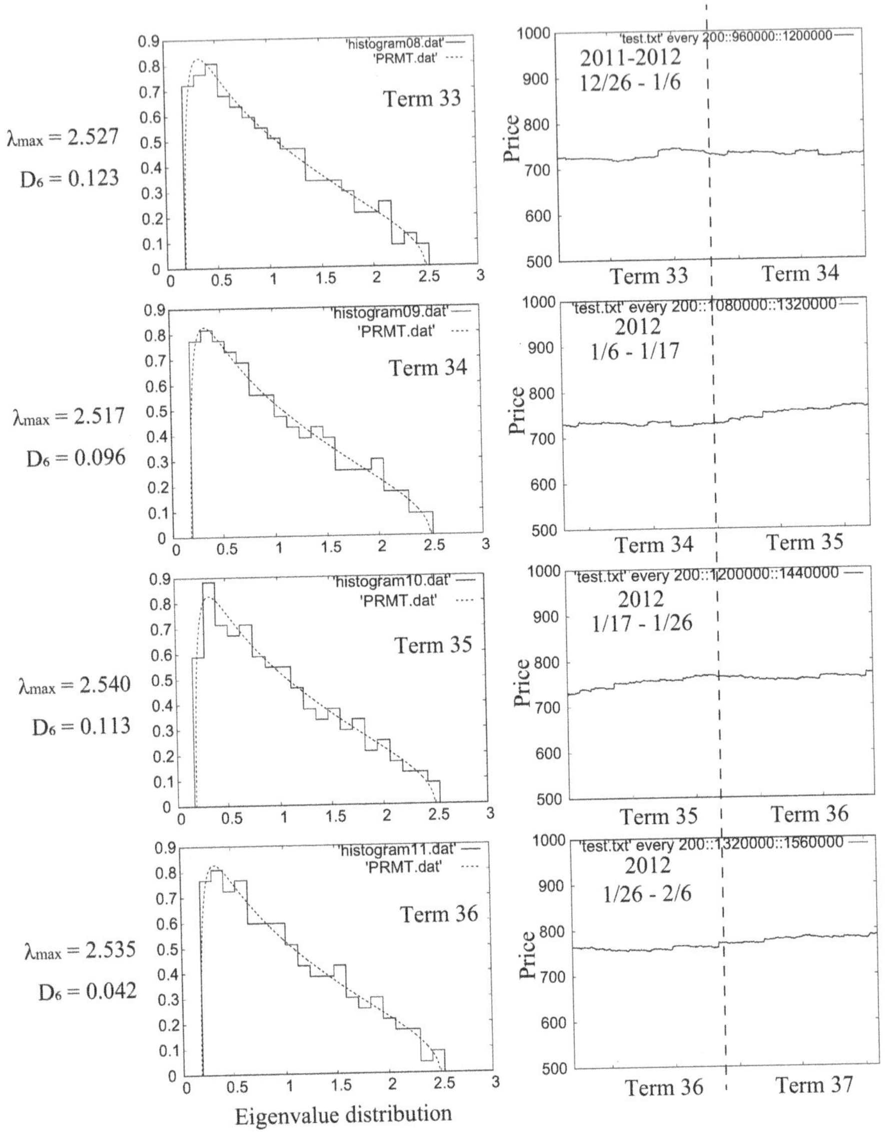

Fig. D.9 Output results for Term 33–36 are shown; λ_{max}, D_6, Eigenvalue Distribution, and Daily index

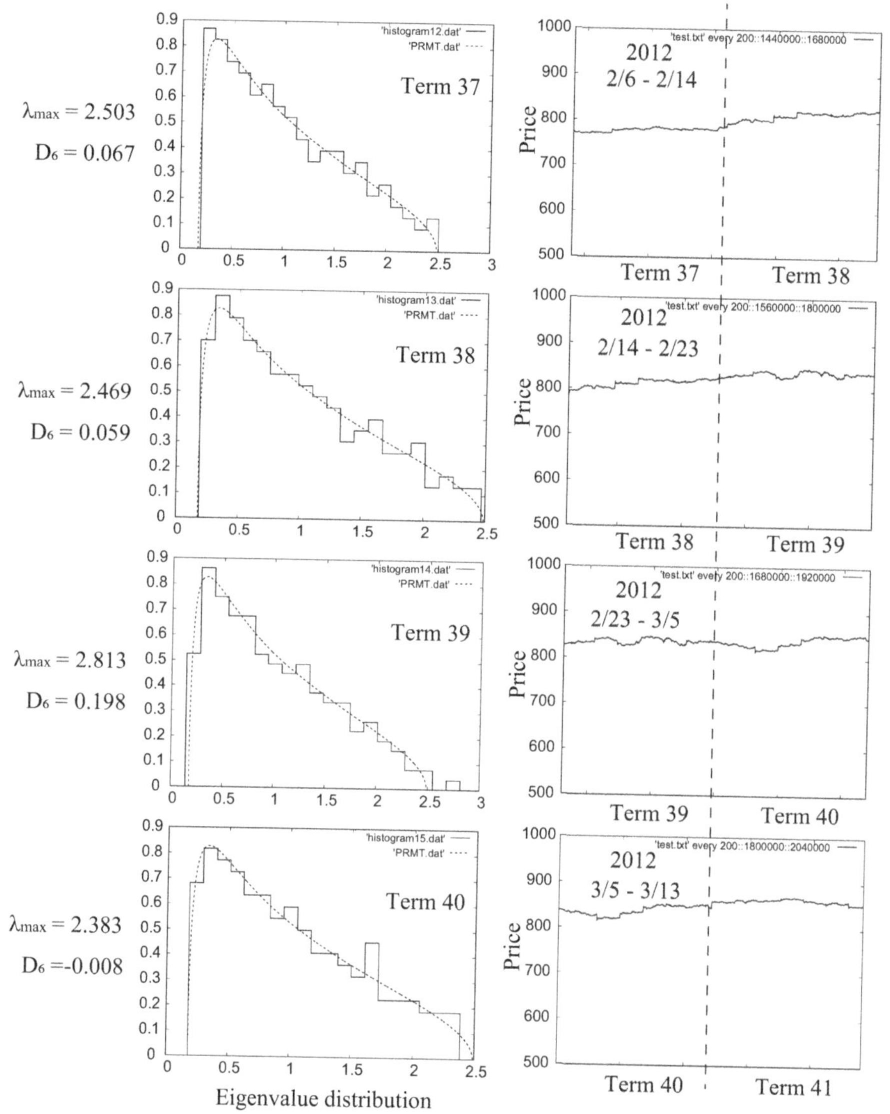

Fig. D.10 Output results for Term 37–40 are shown; λ_{max}, D_6, Eigenvalue Distribution, and Daily index

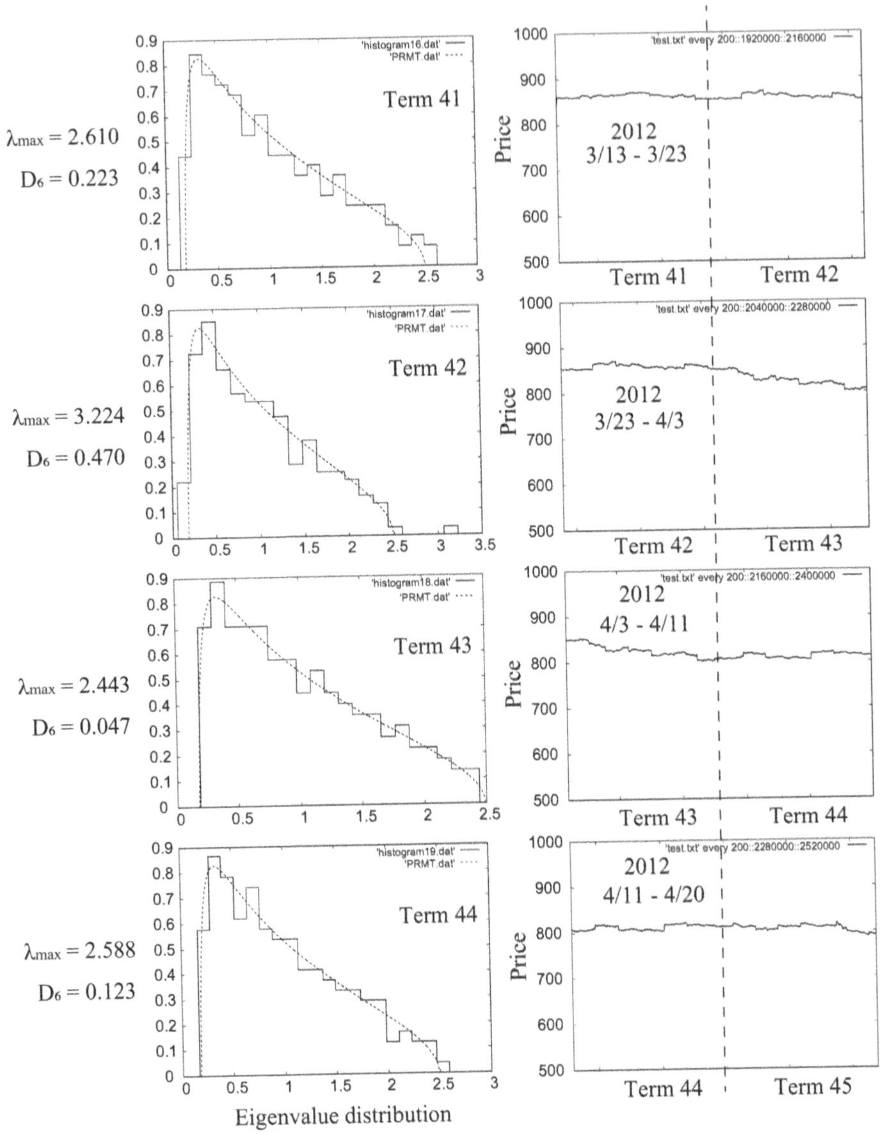

Fig. D.11 Output results for Term 41–44 are shown; λ_{max}, D_6, Eigenvalue Distribution, and Daily index

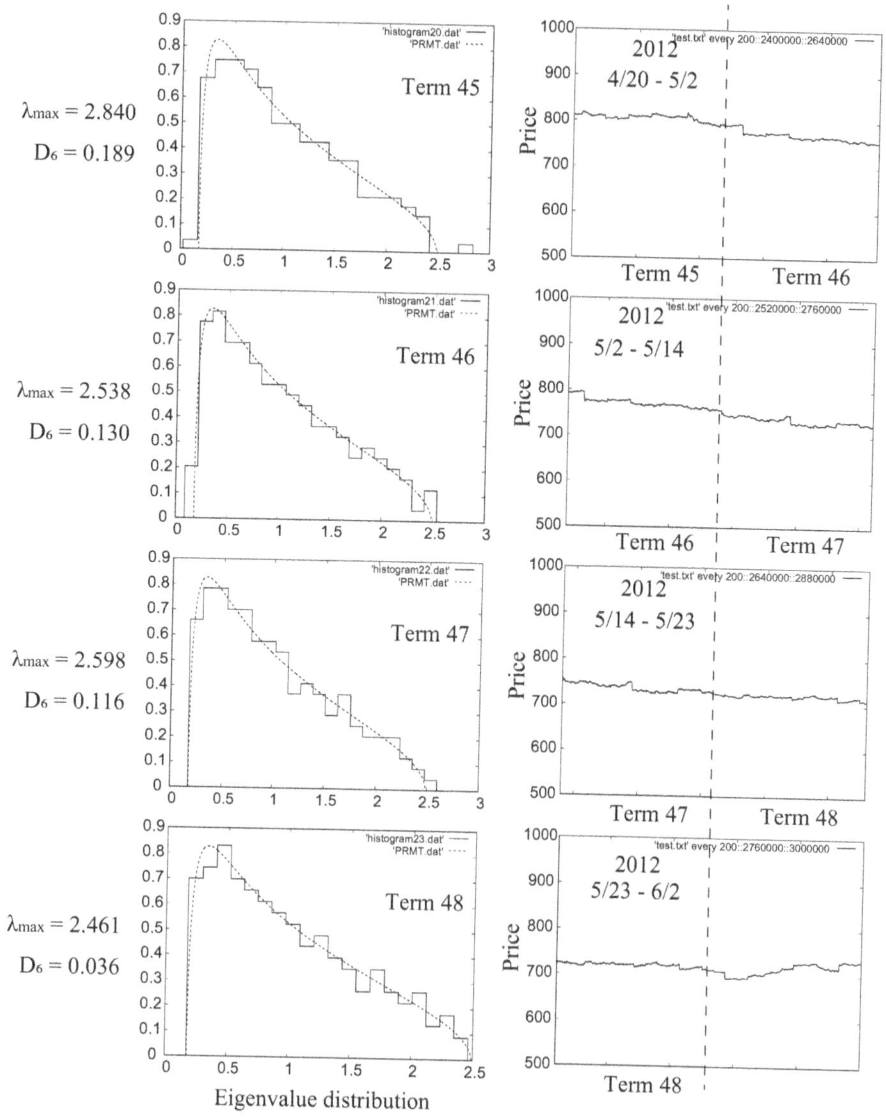

Fig. D.12 Output results for Term 45–48 are shown; λ_{max}, D_6, Eigenvalue Distribution, and Daily index

Appendix E
RMT-test Applied on TOPIXcore30 Index Time Series in 2014

The Result of RMT-test analysis using 1-s price time series of TOPIXcore30 in 2014 is shown, with $Q = 3$, $N = 200$. This is a period when the deviation from the random walk is rather large. Although a slight signal of dropping the randomness is observed in Terms 25 and 26 in Fig. E.7, corresponding to the event shown in Fig. 5.20 and Table 5.14, the rest of the period does not show any remarkable result as can be seen in Figs. E.1, E.6, E.8, and E.9.

M. Tanaka-Yamawaki, Y. Ikura, *Principal Component Analysis and Randomness Test for Big Data Analysis*, Evolutionary Economics and Social Complexity Science 25, https://doi.org/10.1007/978-981-19-3967-9

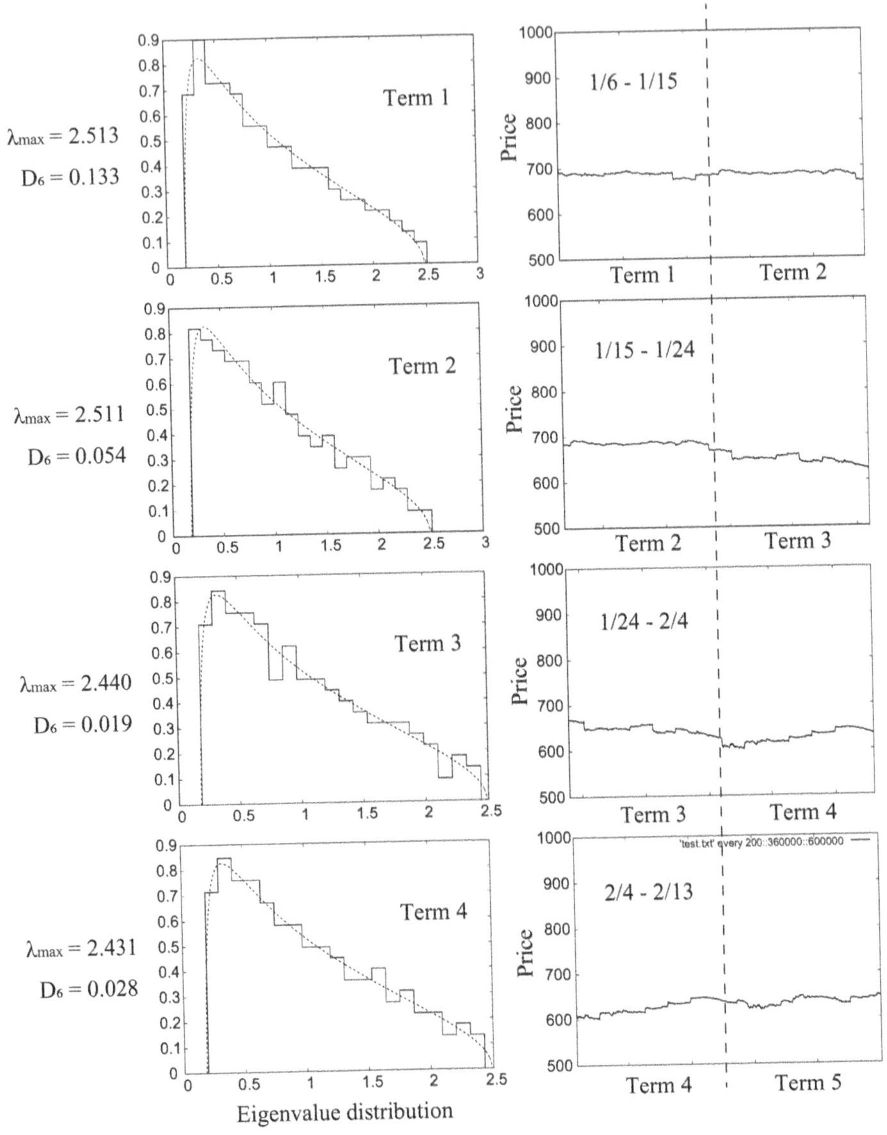

Fig. E.1 Output results for Term 1–4 are shown; λ_{max}, D_6, Eigenvalue Distribution, and Daily index

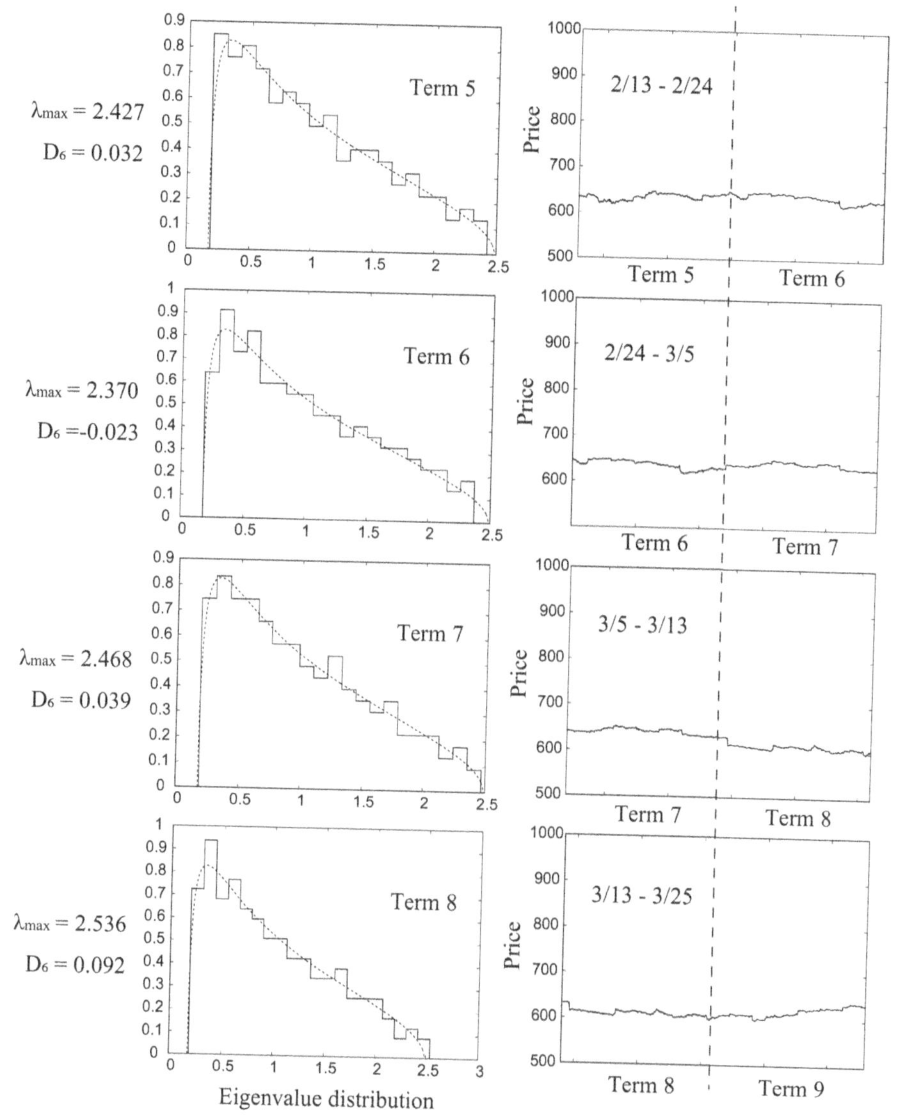

Fig. E.2 Output results for Term 5–8 are shown; λ_{max}, D_6, Eigenvalue Distribution, and Daily index

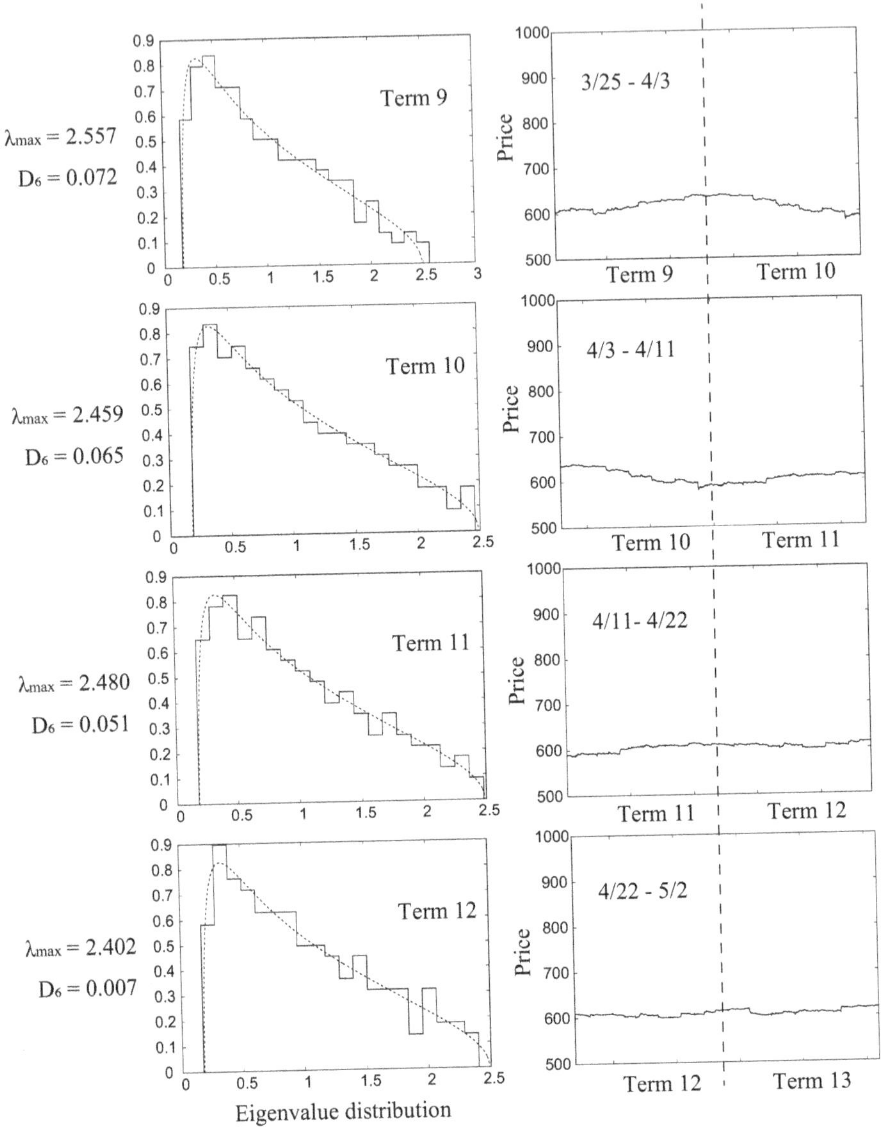

Fig. E.3 Output results for Term 9–12 are shown; λ_{max}, D_6, Eigenvalue Distribution, and Daily index

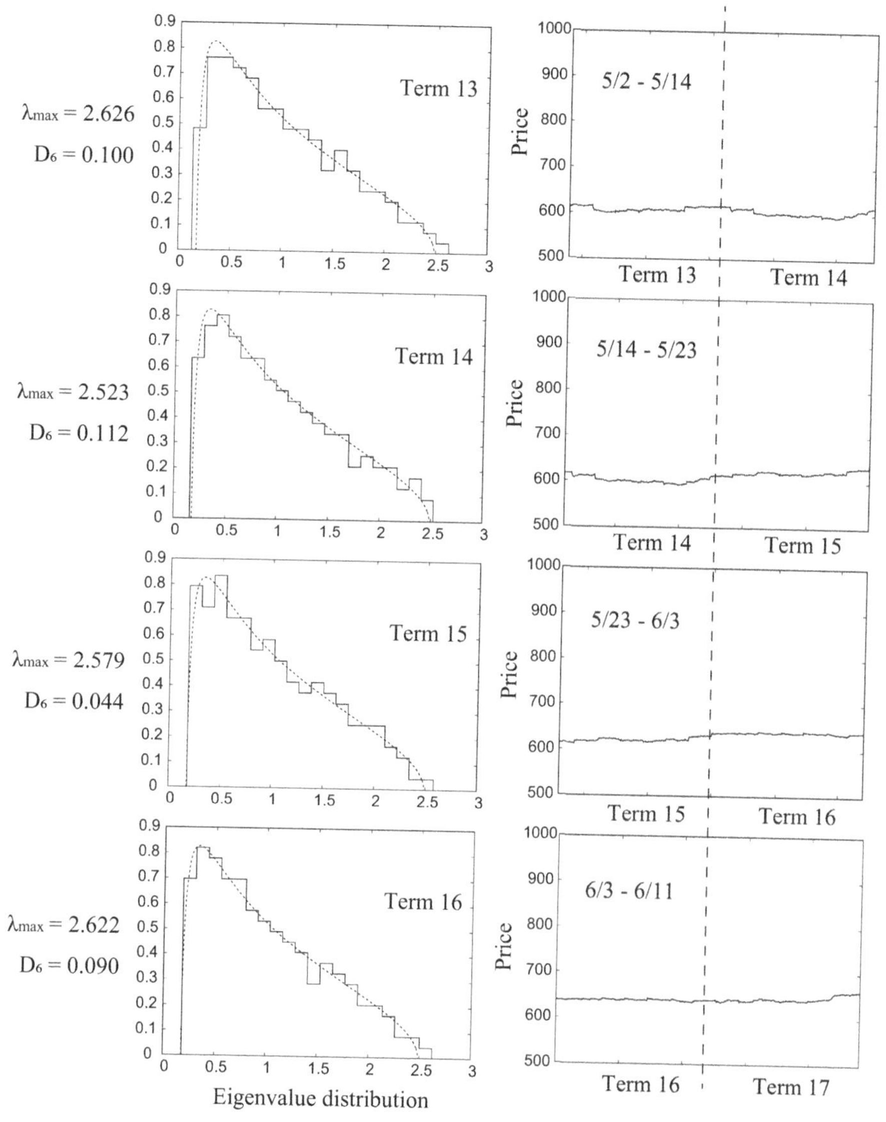

Fig. E.4 Output results for Term 13–16 are shown; λ_{max}, D_6, Eigenvalue Distribution, and Daily index

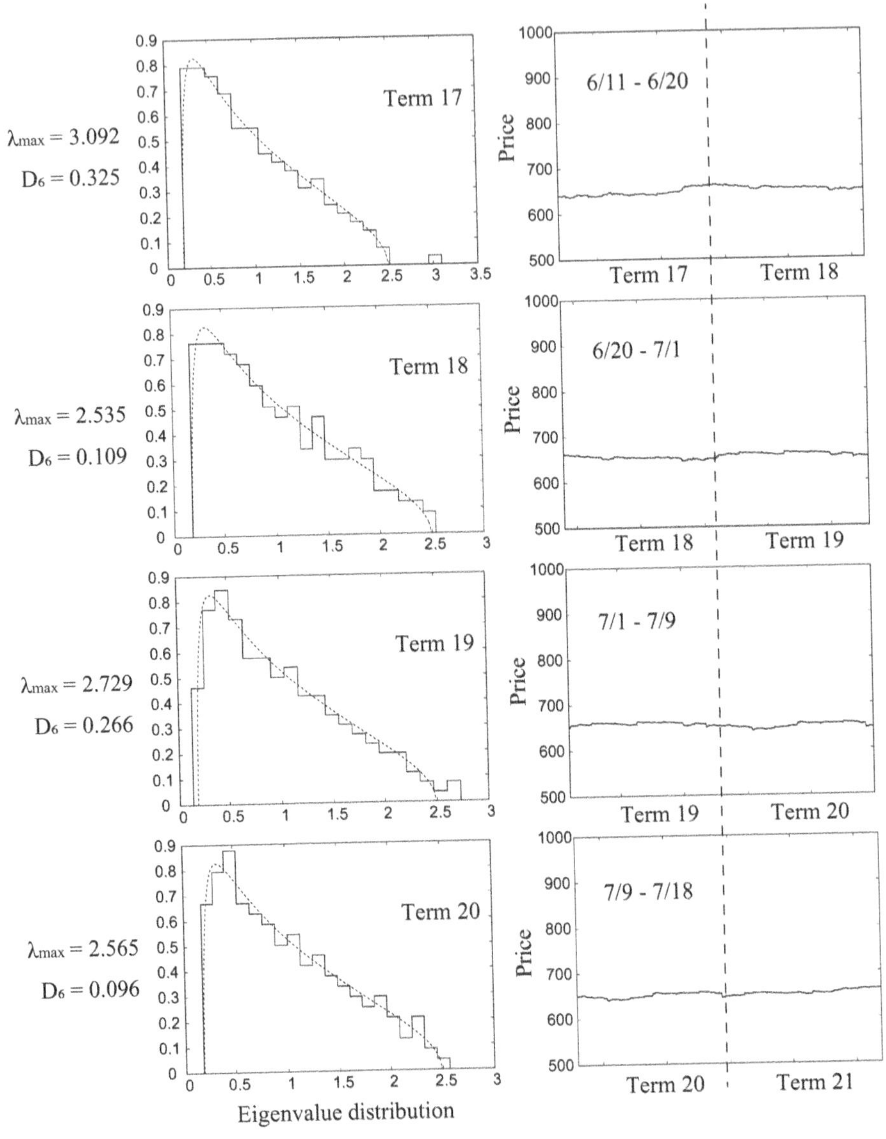

Fig. E.5 Output results for Term 17–20 are shown; λ_{max}, D_6, Eigenvalue Distribution, and Daily index

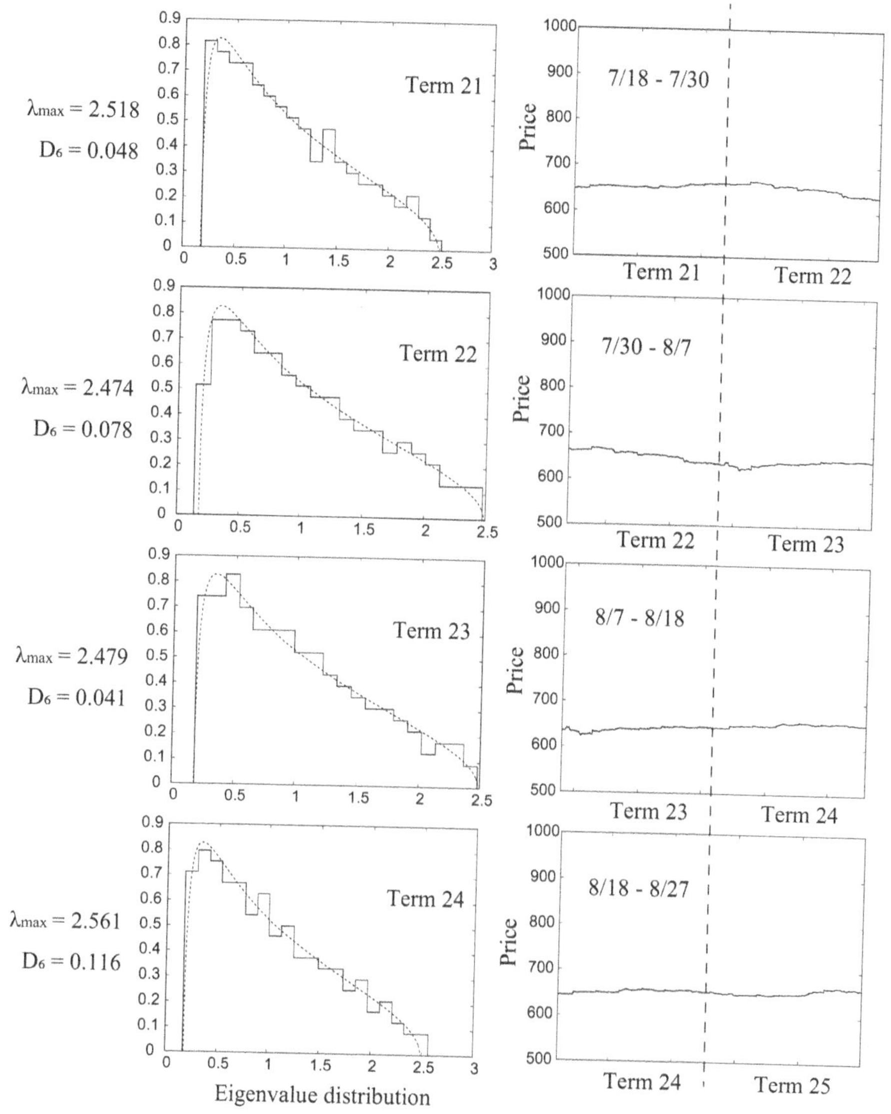

Fig. E.6 Output results for Term 21–24 are shown; λ_{max}, D_6, Eigenvalue Distribution, and Daily index

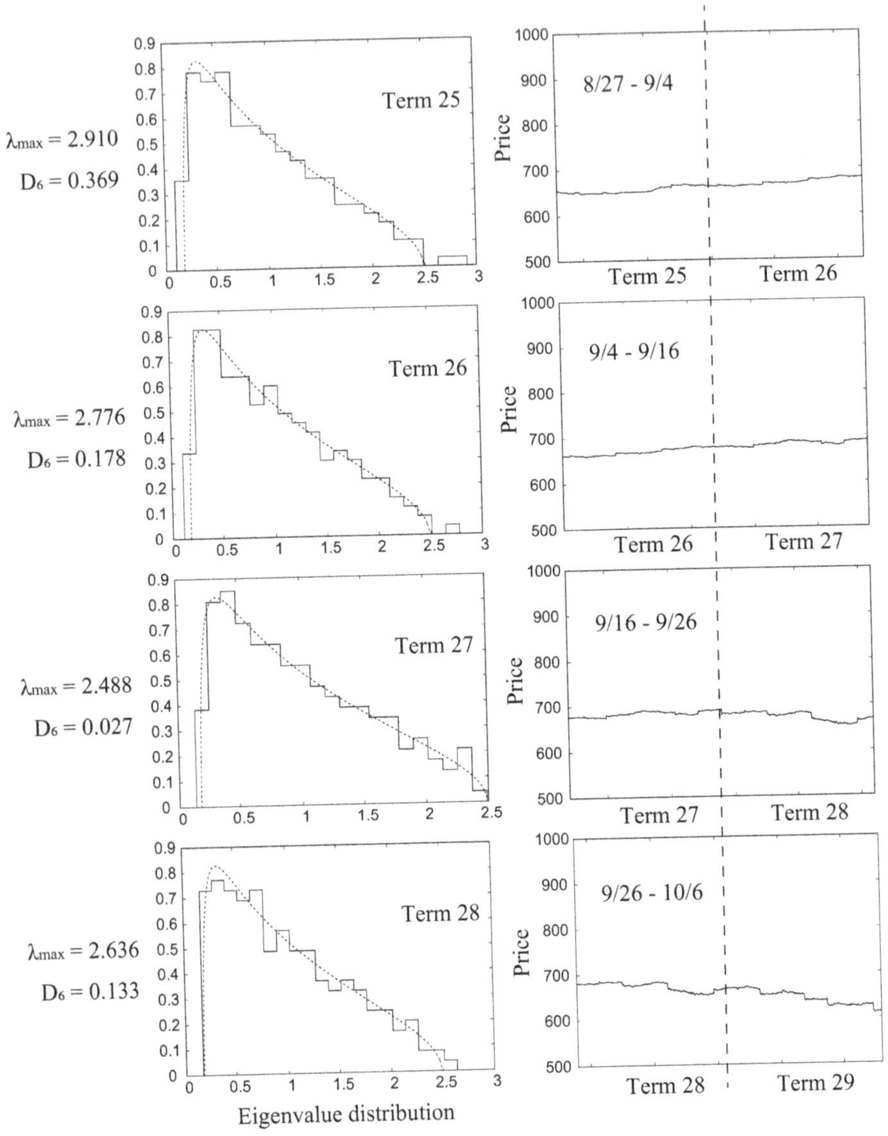

Fig. E.7 Output results for Term 25–28 are shown; λ_{max}, D_6, Eigenvalue Distribution, and Daily index

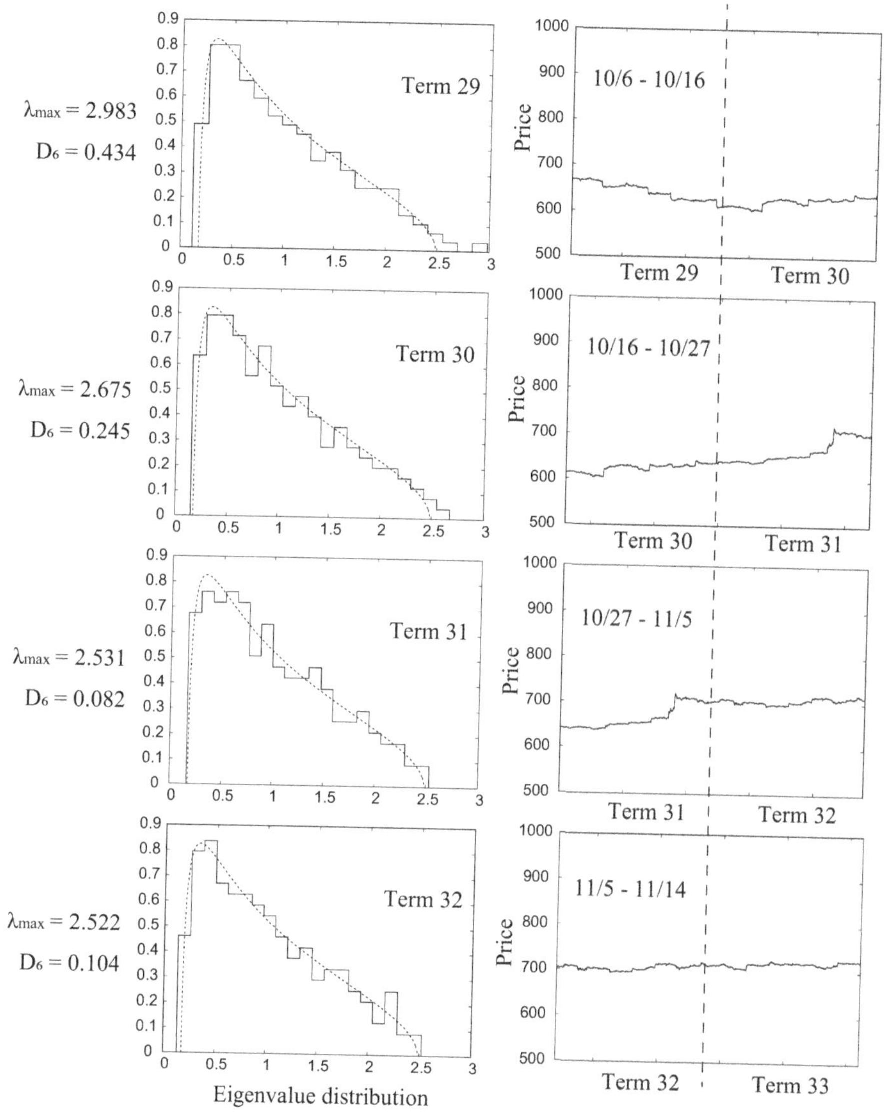

Fig. E.8 Output results for Term 29–32 are shown; λ_{max}, D_6, Eigenvalue Distribution, and Daily index

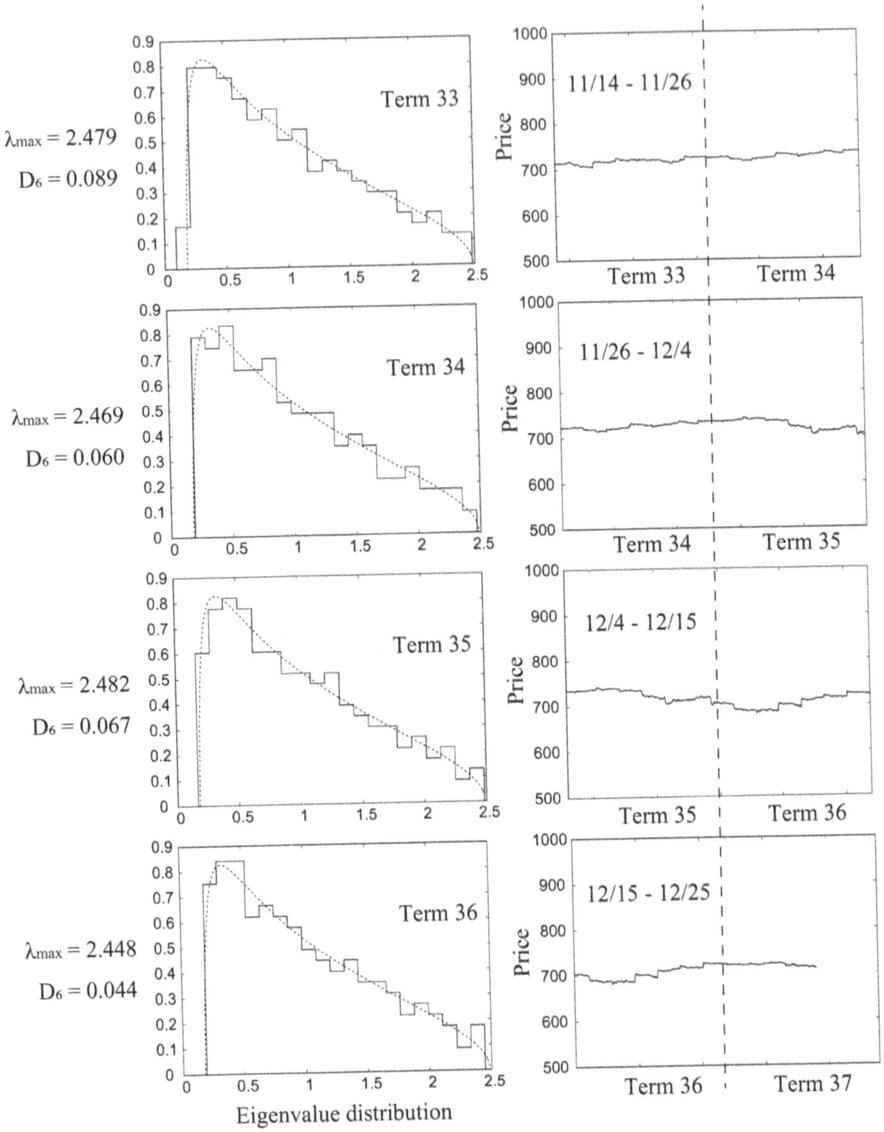

Fig. E.9 Output results for Term 33–36 are shown; λ_{max}, D_6, Eigenvalue Distribution, and Daily index

Bibliography

Wigner E (1955) Characteristic vectors of bordered matrices with infinite dimensions. Ann Math 62:548–564

Mehta ML (2004) Random matrices, 3rd edn. Elsevier, Amsterdam

Edelman A, Rao R (2005) Random matrix theory. Acta Numerica 14:233–297

Bahcall SR (1996) Random matrix model for superconductors in a magnetic field. Phys Rev Lett 77:5276–5279

Beenakker CWJ (1997) Random matrix theory of a quantum transport. Rev Mod Phys 69:731–808

Franchini F, Kravtsov VE (2009) Horizon in random matrix theory, the hawking radiation, and flow of cold atoms. Phys Rev Lett 103:166401

Peyrache A, Benchenane K, Khamassi M, Wiener SI, Battaglia FP (2010) Principal component analysis of ensemble recordings reveals cell assemblies at high temporal resolution. J Comput Neurosci 29:309–325

Park SK, Miller KW (1988) Random number generators: good ones are hard to find. Commun ACM 31:1192–1201

Tanaka-Yamawaki M, Yamanaka M, Ikura YS (2018) Statistical distribution of the sub-second price fluctuation in the latest arrowhead market. Procedia Comput Sci 126:1029–1036

Tanaka-Yamawaki M, Yamanaka M (2019) Market efficiency is truly enhanced in sub-second trading market. Procedia Comput Sci 159:544–551

Laloux L, Cizeau P, Bouchaud J-P, Potters M (1999) Noise dressing of financial correlation matrices. Phys Rev Lett 83:1467–1470

Plerou V, Gopikrishnan P, Rosenow B, Amaral LAN, Stanley HE (1999) Universal and non-universal properties of cross-correlations in financial time series. Phys Rev Lett 83:1471–1474

Bouchaud JP, Potters M (2000) Theory of Financial Risk. Cambridge University Press, Cambridge

Plerou V, Gopikrishnan P, Rosenow B, Amaral LAN, Stanley HE (2002) Random matrix approach to cross correlations in financial data. Phys. Rev. E 65:066126

Tanaka-Yamawaki M, Kido T, Yamamoto A (2010a) Applying random matrix theory to extract principal components of intra-day stock price correlations. In: Proceedings of the 4th International Conference on New Trends in Information Science and Service Science (NISS2010), vol I. IEEE, New York, pp 201–205

Tanaka-Yamawaki M, Kido T, Yamamoto A (2010b) Extracting principal components from pseudo-random data by using random matrix theory. Lect Notes Comput Sci 6278:602–611

Tanaka-Yamawaki M, Kido T, Yamamoto A (2011a) Cross correlation of intra-day stock prices in comparison to random matrix theory. Intell Inf Manag 3:65–70

M. Tanaka-Yamawaki, Y. Ikura, *Principal Component Analysis and Randomness Test for Big Data Analysis*, Evolutionary Economics and Social Complexity Science 25, https://doi.org/10.1007/978-981-19-3967-9

Tanaka-Yamawaki M, Kido T, Yamamoto A (2013) Extracting market trends from the cross correlation between stock time series. Communications in Computer and Information Science 246:25–38
Yang X, Itoi R, Tanaka-Yamawaki M (2011) Testing randomness by means of rmt formula, intelligent decision technologies. In: Intelligent Decision Technologies, vol 10. SIST(Springer), New York, pp 589–596
Tanaka-Yamawaki M, Yang X, Itoi R (2012b) Moment approach for quantitative evaluation of randomness based on RMT formula. In: Intelligent Decision Technologies, vol 16. SIST(Springer), New York, pp 423–432
Yang X, Itoi R, Tanaka-Yamawaki M (2012) Testing randomness by means of random matrix theory. Prog Theor Phys Suppl 194:73–83
Knuth DE (1997) The art of computer programming: seminumerical algorithms, vol 2, 3rd edn. Addison-Wesley, Reading
Matsumoto M, Nishimura T (1998) Mersenne twister: a 623-dimensionally equidistributed uniform pseudo-random number generator. ACM Trans. Model. Comput. Simul. 8:3–30
Tamura Y (2010) Random number library. http://random.ism.ac.jp/random
Aoyama H, Fujiwara Y, Ikeda Y, Iyetomi H, Souma W (2010) Econophysics and companies. Cambrisdge University Press, Cambrisdge
Sengupta AM, Mitra PP (1999) Distribution of singular values for some random matrices. Phys Rev E 60:3389–3392
Tanaka-Yamawaki M, Ikura YS (2020) Tracking the annual changes of statistical property of a stock index in the arrowhead market. Procedia Comput Sci 176:1626–1632
Tanaka-Yamawaki M, Ikura YS (2022) Possible relationship of the randomness and the stock performance. In: Aruka (ed) Digital designs for money, markets and social dilemmas. Evolutionary Economics and Social Complexity Science, vol 28, chapter 14. Springer, Singapore
Perron O (1907) Zur theorie der matrices. Math Ann 64(2):248–263
Frobenius G (1912) Ueber matrizen aus nicht negativen elementen. Sitzung sberichte der K'oniglich Preussischen Akademie der Wissenschaften 26:456–477
Marchenko VA, Pastur LA (1967) Distribution of eigenvalues for some sets of random matrices. Mathematics of the USSR-Sbornik 1:457–483
Yang X, Mikamori Y, Tanaka-Yamawaki M (2013a) Predicting the security levels of stock investment by using the RMT-test. Prog Theor Phys Suppl 22:1172–1181
Black J, Cochran M, Highland T (2008) A study of the md5 attacks: insights and improvements. http://www.cs.colorado.edu/jrblack/papers/md5efull.pdf
Locktyukhin M, Farrel K (2010) Improving the performance of the secure hash algorithm (SHA-1). http://software.intel.com/en-us/articles/improvingthe-performance-of-the-secure-hash-algorithm-1/
Bachelier L (1900) Théorie de la spéculation. Annales scientifiques de l'École Normale Supérieure 17:21–86
Mantegna RN, Stanley HE (1995) Scaling behavior in the dynamics of an economic index. Nature 376:46–49
Mantegna RN, Stanley HE (2000) *An Introduction to Econophysics: Correlations and Complexity in Finance*. Cambridge University Press, Cambridge
Chandrasekhar S (1943) Stochastic problems in physics and astronomy. Rev Mod Phys 15(1):1–89
Kahneman D, Tversky A (1979) Prospect theory: an analysis of decision under risk. Econometrica 47(2):263–292

The manufacturer's authorised representative in the EU is Springer Nature Customer Service Centre GmbH, Europaplatz 3, 69115 Heidelberg, Germany. If you have any concerns regarding our products, please contact ProductSafety@springernature.com

Printed and bound by CPI Group (UK) Ltd, Croydon, CR0 4YY
02/10/2026
02226885-0005